SPINOFF

Office of the Chief Technologist

2010

On the cover: During the STS-128 space shuttle mission, a space-walking astronaut took this photograph of a portion of the International Space Station (ISS) flying high above Earth's glowing horizon. This year marks the 10th anniversary of the ISS, and is also a year of beneficial spinoff technologies, as highlighted by the smaller images lined up against the backdrop of space.

Spinoff Program Office
NASA Center for AeroSpace Information

Daniel Lockney, *Editor*
Bo Schwerin, *Senior Writer*
Lisa Rademakers, *Writer*
John Jones, *Graphic Designer*
Deborah Drumheller, *Publications Specialist*

For sale by the Superintendent of Documents, U.S. Government Printing Office
Internet: bookstore.gpo.gov   Phone: toll free (866) 512-1800; DC area (202) 512-1800
Fax: (202) 512-2104 Mail: Stop IDCC, Washington, DC 20402-0001

ISBN 978-0-16-086326-4

Table of Contents

7   Foreword
9   Introduction
10 International Space Station Spinoffs
20 Executive Summary
30 NASA Technologies Enhance Our Lives on Earth
32 NASA Partnerships Across the Nation
34 NASA Technologies Benefiting Society

Health and Medicine

Burnishing Techniques Strengthen Hip Implants 38
Signal Processing Methods Monitor Cranial Pressure 40
Ultraviolet-Blocking Lenses Protect, Enhance Vision 42
Hyperspectral Systems Increase Imaging Capabilities 44

Transportation

Programs Model the Future of Air Traffic Management 48
Tail Rotor Airfoils Stabilize Helicopters, Reduce Noise 50
Personal Aircraft Point to the Future of Transportation 52
Ducted Fan Designs Lead to Potential New Vehicles 56
Winglets Save Billions of Dollars in Fuel Costs 58
Sensor Systems Collect Critical Aerodynamics Data 60
Coatings Extend Life of Engines and Infrastructure 62

Public Safety

Radiometers Optimize Local Weather Prediction 66

Energy-Efficient Systems Eliminate Icing Danger for UAVs 68

Rocket-Powered Parachutes Rescue Entire Planes 70

Technologies Advance UAVs for Science, Military 72

Inflatable Antennas Support Emergency Communication 74

Smart Sensors Assess Structural Health 76

Hand-Held Devices Detect Explosives and Chemical Agents 78

Terahertz Tools Advance Imaging for Security, Industry 80

Consumer Goods

LED Systems Target Plant Growth 84

Aerogels Insulate Against Extreme Temperatures 86

Image Sensors Enhance Camera Technologies 90

Lightweight Material Patches Allow for Quick Repairs 92

Nanomaterials Transform Hairstyling Tools 94

Do-It-Yourself Additives Recharge Auto Air Conditioning 96

Spinoff 2010   *Table of Contents*   3

Environmental Resources

Systems Analyze Water Quality in Real Time 100
Compact Radiometers Expand Climate Knowledge 102
Energy Servers Deliver Clean, Affordable Power 104
Solutions Remediate Contaminated Groundwater 106
Bacteria Provide Cleanup of Oil Spills, Wastewater 108
Reflective Coatings Protect People and Animals 110

Computer Technology

Innovative Techniques Simplify Vibration Analysis 114
Modeling Tools Predict Flow in Fluid Dynamics 116
Verification Tools Secure Online Shopping, Banking 118
Toolsets Maintain Health of Complex Systems 120
Framework Resources Multiply Computing Power 122
Tools Automate Spacecraft Testing, Operation 124
GPS Software Packages Deliver Positioning Solutions 126
Solid-State Recorders Enhance Scientific Data Collection 128
Computer Models Simulate Fine Particle Dispersion 130

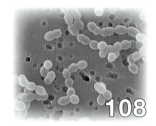

Industrial Productivity

Composite Sandwich Technologies Lighten Components 134

Cameras Reveal Elements in the Short Wave Infrared 136

Deformable Mirrors Correct Optical Distortions 138

Stitching Techniques Advance Optics Manufacturing 140

Compact, Robust Chips Integrate Optical Functions 142

Fuel Cell Stations Automate Processes, Catalyst Testing 144

Onboard Systems Record Unique Videos of Space Missions 146

Space Research Results Purify Semiconductor Materials 148

Toolkits Control Motion of Complex Robotics 150

152 Aeronautics and Space Activities

178 Education News

188 Partnership News

NASA Technology Award Winners 204

210 Office of the Chief Technologist

Denotes that *R&D Magazine* has awarded the technology with its "R&D 100" award.

Indicates that the Space Foundation has inducted the technology into the Space Technology Hall of Fame.

Signifies that the technology has been named as a NASA "Invention of the Year."

Designates that the software has been named as a NASA "Software of the Year."

For a list of all *Spinoff* award winners since publication began in 1976, please see page 204.

146

148

168

187

194

DISCLAIMER: While NASA does not manufacture, market, or sell commercial products, many commercial products are derived from NASA technology. Many NASA-originated technologies are adapted by private industry for use by consumers like you. Spinoff developments highlighted in this publication are based on information provided by individual and private industry users of NASA-originated aerospace technology who acknowledge that such technology contributed wholly or in part to development of the product or process described. NASA cannot accept responsibility or liability for the misinterpretation or misrepresentation of the enclosed information provided by these third-party users. Publication herein does not constitute NASA endorsement of the product or process, nor confirmation of manufacturers' performance claims related to any particular spinoff development.

Foreword

Since NASA's inception in 1958, the Agency has been charged with ensuring its research and development activities can be shared and applied beyond the space community. NASA spinoffs are one result. These are the technologies and products the Agency has successfully shared with industry, which in turn has developed and refined them for many benefits, including medical advances, a cleaner environment, safer households, and more convenience in our daily lives.

NASA is always finding ways to bring the benefits of space exploration back to Earth. You will find NASA in the average household in many ways. It might not be obvious that the air purifier in a refrigerator or a handheld cordless vacuum came about as a result of space missions, but they did.

In fact, NASA's research and development has had a major and positive impact on public welfare. Technologies we can trace to the earliest days of the Space Program have improved water purification systems. NASA innovations have brought us advanced home insulation and fire-resistant fabrics used by firefighters and soldiers. NASA technology can even be found in infant formula and modern semi truck design. The list is expansive. As a Nation, we have received a significant return on our investment in space, and we have advanced our capabilities in many areas thanks to this ongoing flow of NASA ideas and technology.

NASA defines a spinoff as a commercially available product, service, or process that takes NASA-related technology and brings it to a broader audience. While the original purposes were mission-related, the technologies now are filling needs in everyday life. From robotics-based nutrition programs to better swimsuits and other sports equipment, NASA innovation has advanced our standard of living.

Since 1976, NASA has been documenting these spinoffs. It is an interesting and varied history. We are pleased to present to you this annual report on our latest innovations and ways we are inspiring people beyond our science and exploration missions.

We see it this way: NASA provides a spark of inspiration, a seed of technology, and then industry carries the ball forward and transforms it into something the general public can use. Water purification technology originally developed for the International Space Station, for instance, can bring clean water to people in remote areas where there is none. The things we learn in the coming decade on the station and in the development of new systems for reaching deep space will have far-reaching benefits. In this sense, spinoffs are representative of the new era of global exploration.

The new age of exploration will require innovative and robust technology development. NASA will continue to pursue fresh innovations and partnerships, and the Agency's renewed commitment to research and development will bring benefits to people everywhere in the decades to come. We truly can make life better for everyone on the planet.

Charles F. Bolden, Jr.
Administrator
National Aeronautics and Space Administration

For over 50 years, NASA has created new technologies with direct benefit to the private sector, supporting global competition and the economy. The resulting commercialization has contributed to products and services in the fields of health and medicine, transportation, public safety, consumer goods, environmental resources, computer technology, and industry. Since 1976, NASA has featured these technologies in its *Spinoff* publication.

Introduction

As a research and development agency, NASA plays a vital role in America's innovation engine and, as such, its future economic prosperity and security. The President's FY 2011 budget request for NASA is part of a larger national research and development effort in science, technology, and innovation that will lead to new products and services, new business and industries, and high-quality, sustainable jobs. NASA's new technology and innovation investments are required to enable new approaches to NASA's current aeronautics, science, and exploration missions and allow the Agency to pursue entirely new missions including sending humans into deep space to compelling destinations such as near-Earth asteroids and Mars. In tandem with these technology investments, NASA will continue to ensure an American presence in space aboard the International Space Station and empower a robust and competitive American commercial space program.

NASA's new Space Technology programs will foster cutting-edge, competitively sponsored research and technology development efforts in academia, industry, the NASA Centers, and other government entities, rebuilding our core competencies and allowing innovative technological solutions to today's challenges. These new space technology investments will create a more vital and productive aerospace industry and address broader national needs, such as energy, health and wellness, and national security.

NASA's technology, expertise, and facilities are already a valuable national asset with a long history of providing innovation and inspiration for the good of the American public. Since its first days, NASA has nurtured partnerships with the private sector to facilitate the transfer of its technologies to improve the lives of Americans and people around the world.

Each year since 1976, NASA has chronicled some of the best examples of this successful technology transfer in its premier journal, *Spinoff*. The remarkable outcomes of these partnerships have reached throughout the economy and around the globe, as the resulting commercial products contributed to the development of services and technologies in the fields of health and medicine, transportation, public safety, consumer goods, environmental resources, computer technology, and industry.

This year is no exception, as this latest edition of *Spinoff* reveals a wide range of public benefits. In the following pages, you will find these noteworthy examples and more:

- Light sensors invented by NASA researchers provide imaging capabilities for digital cameras, Web cameras, automotive cameras, and one of every three cell phone cameras on the planet.

- NASA funding supported the development of a whole aircraft parachute system that is now standard equipment on many of the world's top-selling aircraft and has saved 246 lives to date.

- Fuel cell technology originally devised for generating oxygen and fuel on Mars has been adapted to generate clean energy on Earth, providing an environmentally friendly, scalable power source for a host of Fortune 500 businesses.

- Drag-reducing winglet technology—advanced and proven by NASA researchers—now features on aircraft around the world, saving the airline industry billions of dollars in fuel costs and significantly reducing carbon emissions.

- Bacteria isolated for use in water-purifying technology for the International Space Station is providing a safe, environmentally sound method for oil spill cleanup and for cleansing municipal and industrial wastewater.

These are just a few of the positive stories of NASA technologies leaving the laboratories to improve life on Earth and just some of the over 1,700 examples that have been recorded in *Spinoff* over the years.

Dr. Robert D. Braun
Chief Technologist
National Aeronautics and Space Administration

NASA's new technology investments represent an important aspect of our overall national investment in research, technology, and innovation, designed to stimulate our economy, create new inventions and capabilities, and increase our global economic competitiveness. We predict that as the Agency continues to push technological boundaries and follow in its proud tradition of doing things that have never been done before—and sometimes things that had never been thought possible—it will carry on producing new and exciting technologies that will, no doubt, further improve our lives back here on Earth.

While NASA's research does indeed provide valuable scientific outcomes and clear public benefits, there is perhaps another "spinoff" even more lucrative than the benefits we see from these industry partnerships. As President Obama noted in an address to the National Academy of Sciences, in which he invoked the many tangible benefits of the Nation's investment in the Apollo Program, "The enormous investment of that era—in science and technology, in education and research funding—produced a great outpouring of curiosity and creativity, the benefits of which have been incalculable."

NASA's investment in new technologies is also an investment in our country's future. Today's children will be inspired by NASA's bold new vision, and our new technology and innovation emphasis will create a pipeline of young engineers, scientists, and mathematicians to serve our future national needs, inspiring wonder in a new generation, sparking passions, and launching careers.

International Space Station Spinoffs

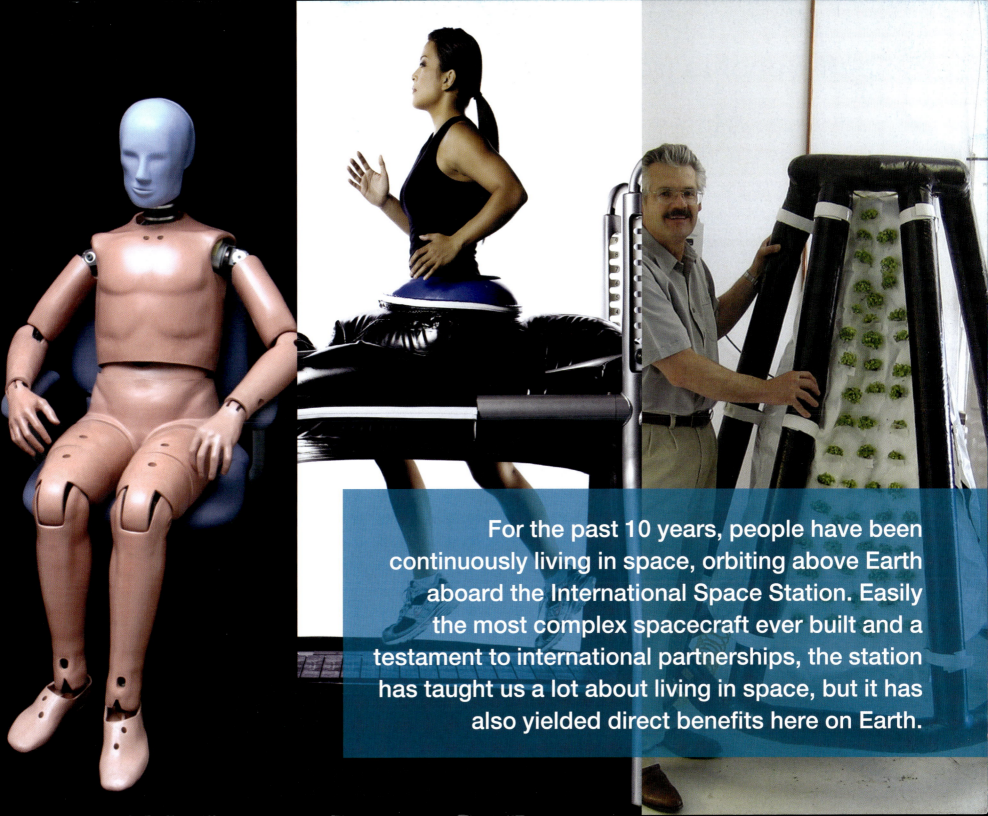

For the past 10 years, people have been continuously living in space, orbiting above Earth aboard the International Space Station. Easily the most complex spacecraft ever built and a testament to international partnerships, the station has taught us a lot about living in space, but it has also yielded direct benefits here on Earth.

Spinoff Benefits from the International Space Station

On many nights, you can see the International Space Station (ISS) whizzing by overhead. You have to know just where to look, though, and you have to be quick. It only takes the station about 90 minutes to orbit the Earth, so it may only be visible for a few minutes at a time. As it passes over your head, know this: The ISS is marking its 10th anniversary of continuous habitation in orbit this year. That means people have been living in space now for 10 years, approximately 220 miles above us, passing overhead several times a day.

The floating laboratory, a cooperative effort between 15 international partners and 5 space agencies, now supports a multicultural crew of 6. The 6 currently

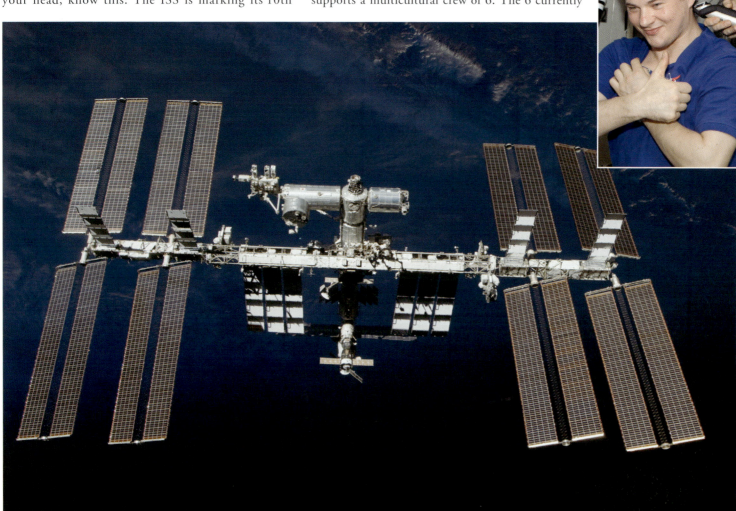

NASA astronaut Tim Kopra (above) trims Russian cosmonaut Roman Romanenko's hair in the Destiny Laboratory of the International Space Station (ISS). NASA astronaut Nicole Stott looks on. Kopra used hair clippers fashioned with a vacuum device to garner freshly cut hair. The ISS (left) is featured in this image photographed by an STS-130 crewmember on Space Shuttle Endeavour after the station and shuttle began their separation.

Crewmembers onboard the ISS share a meal in the Unity Node. Pictured from the left (bottom) are NASA astronauts Rick Sturckow, STS-128 commander; Tim Kopra and Jose Hernandez, both STS-128 mission specialists; along with Kevin Ford, STS-128 pilot; and John "Danny" Olivas (mostly out of frame at right), STS-128 mission specialist. Pictured from the left (top, partially out of frame) are NASA astronaut Nicole Stott and Canadian Space Agency astronaut Robert Thirsk, both Expedition 20 flight engineers; along with NASA astronaut Patrick Forrester, STS-128 mission specialist.

Astronaut Nicole Stott, Expedition 20 flight engineer, participates in the STS-128 mission's first session of extravehicular activity as construction and maintenance continue on the ISS.

on-orbit are just some of the nearly 200 astronauts, cosmonauts, and space tourists who have ventured to the station, a spacecraft that over the last decade has grown to a massive 800,000 pounds, with a habitable volume of more than 12,000 cubic feet. Approximately the size of a five-bedroom house, this remarkable testament to human will and engineering prowess uses sophisticated systems to generate solar electricity and recycles much of its of water (nearly 85 percent) and oxygen supply.

The ISS is the most advanced spacecraft ever built, and unlike the space race of the 1960s that culminated with American astronauts landing on the Moon, this is an international partnership, a team effort. In addition to station assembly, the international partner agencies (NASA, the Canadian Space Agency, the European Space Agency, the Japan Aerospace Exploration Agency, and the Russian Federal Space Agency) train and launch crews and provide ground support for the orbiting research facility.

Ten Years in the Making

Construction of the ISS began when the Zarya Control Module was launched atop a Russian rocket from Baikonur Cosmodrome, Kazakhstan, on November 20, 1998. The Zarya module provides battery power, fuel storage, and rendezvous and docking capability for Soyuz

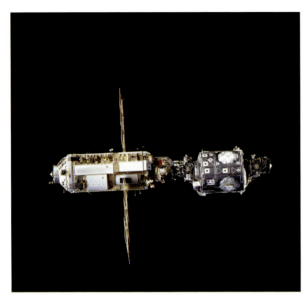

The U.S.-built Unity connecting module and the Russian-built Zarya module are backdropped against the blackness of space in this photograph taken December 1988 from the Space Shuttle Endeavour. After devoting the major portion of its mission time to ready the two docked modules for their ISS roles, the six-member STS-88 crew released the tandem and performed a fly-around survey of the hardware.

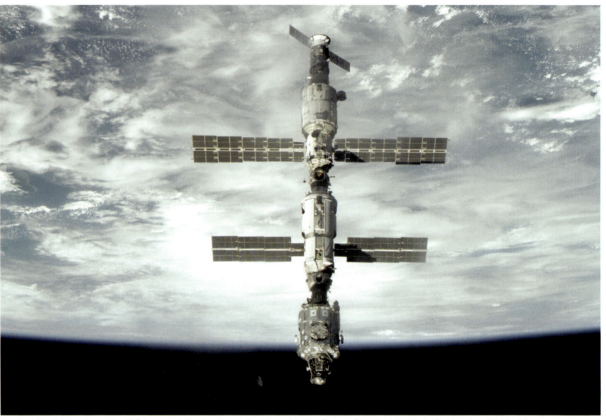

Against Earth's horizon, the ISS is seen following its undocking with the Space Shuttle Atlantis in September 2000. After accomplishing all mission objectives in outfitting the station for the first resident crew, the seven astronauts and cosmonauts undocked and snapped this picture.

and Progress space vehicles. Just a few days later, on December 4, 1998, the U.S.-built Unity node launched aboard Space Shuttle Endeavor. During three spacewalks, the crew connected power and data transmission cables between Unity and Zarya. The Unity node had two pressurized adapters, one of which was permanently affixed to the Russian unit, the other was designed for space station docking. Inside Unity were a series of additional ports and passageways, each with a sign designating where additional modules would be attached.

In October 2000, approximately 2 years after assembly began, the first crew to live on the station launched aboard a Soyuz spacecraft. With the Soyuz capsule docked, the crew, referred to as Expedition 1, had a way to return back home. In March 2001, Space Shuttle Discovery carried an Italian-built component, the Leonardo Multi-Purpose Logistics Module, the first of three large modules designed to serve as moving vans for the station. Approximately 21 feet in length and 15 feet in diameter, these canisters can be carted back and forth between Earth and the station in the cargo bay of the space shuttles, or can serve as workable living space aboard the station. With delivery of the first module came the second crew to live aboard the station, Expedition 2. This was followed by a series of additional shuttle deliveries, including infrastructure for enabling spacewalks, stowage racks and life support systems, racks for experiments, and Canadarm2, a robotic arm that would prove useful for future station assembly. The next handful of missions involved delivery and assembly of the truss systems and solar arrays and the rotation of more crews.

Delivery of large structural components to the station was interrupted for a 3-year period following the loss of Space Shuttle Columbia and her crew after the craft disintegrated during atmospheric reentry in early 2003.

Backdropped by Earth dotted with clouds, this close-up view of the ISS was taken by one of the crewmembers on the Space Shuttle Discovery after undocking in August 2001 after more than a week of joint operations.

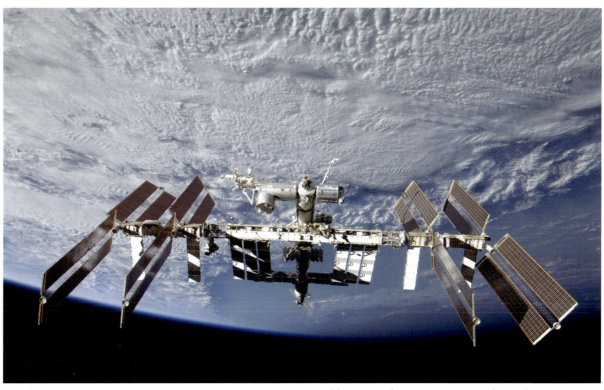

With Earth's horizon and the blackness of space in the background, the ISS is seen in September 2009 from Space Shuttle Discovery as the two spacecraft begin their separation after the STS-128 and Expedition 20 crew concluded 9 days of cooperative work onboard the shuttle and station.

During this time, the Russian Soyuz crafts continued to ferry expedition crews back and forth between Earth and the orbiting laboratory.

It was not until the space shuttles were again flying in 2005 that construction of the orbiting laboratory truly resumed. With the return to flight came designation of the U.S. portion of the ISS as a national laboratory by Congress as part of the 2005 NASA Authorization Act, signaling a renewed dedication to full utilization of this structure for science projects. Additional truss segments followed, as did several large solar arrays. These were followed by installation of the Harmony Node 2, which in addition to creating extra work space provided couplings for connecting the European Columbus Laboratory and the Japanese Kibo Laboratory.

In 2008, Space Shuttle Endeavour delivered supplies and equipment, including additional crew quarters, exercise equipment, equipment for the regenerative life support system, and spare hardware, inside the Leonardo Multi-Purpose Logistics Module. The 2008 mission also saw delivery of parts for Kibo.

In May 2009, the STS-119 crew of Space Shuttle Discovery delivered and installed the ISS's final, major U.S. truss segment, Starboard 6, and its final pair of power-generating solar arrays. Later that year, astronauts attached the Kibo Japanese Experiment Module Exposed Facility and Experiment Logistics Module Exposed Section, providing a "front porch" for the facility where astronauts could conduct experiments outside of the spacecraft.

NASA then delivered life support and science racks. In November 2009, the space shuttle made its last delivery of crewmembers to the ISS.

In 2009, the ISS Program received the Collier Trophy, considered by many to be the top award in aviation. In 2010 it also received the Aviation Week "Space Laureate Award." Perhaps more significant, though, is that 10 years of assembly is coming to an end, Congress has extended the life of the station, and full-time scientific experimentation will now begin in earnest.

Current research aboard the ISS is steadily progressing, such as experiments to understand the muscular deterioration of astronauts' hearts in the reduced gravity environment, which will add to the understanding of heart function here on Earth, particularly among patients who are confined to beds for long periods of time or wheelchair bound; and experiments to grow and harvest new crops in space that show promise for producing biofuels that can be used as energy sources for future space missions or here on Earth. More on these projects in progress, and additional information about recent missions, expeditions, and assembly can be found in the Space Operations Mission Directorate portion of the Aeronautics and Space Activities chapter, starting on page 172.

 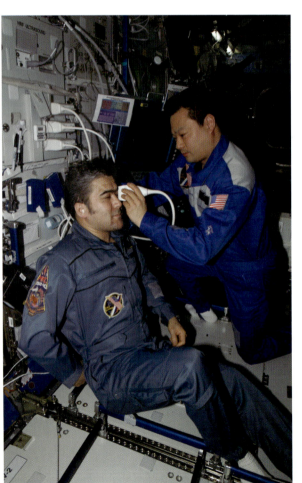

Remote ultrasound procedures provide for medical diagnoses to areas as far-flung as Mount Everest and the ISS—miles from professional medical personnel.

International Space Station Spinoffs

High above Earth, the ISS provides a research platform where nearly 150 experiments are underway. Over the years, more than 400 experiments have been conducted. Taking advantage of the unique environment of microgravity, these experiments cover a wide variety of disciplines, including human life sciences, biological science, human physiology, physical and material science, and Earth and space science. These experiments are developing new ways to fight disease, advances in understanding food-borne illnesses, growing crops for alternative energy usage, and the development of superior materials for use in both space and on Earth.

Many of the technologies developed for the ISS have resulted in practical, tangible benefits that we find here on Earth.

Bioreactors Advance Disease Treatments

A NASA device used to cultivate healthy cell tissues for space station and Earth experiments is now enhancing medical research. Treatments developed using bioreactor-grown cells may be used to counter conditions like heart disease, diabetes, and sickle cell anemia (*Spinoff* 2009).

Image-Capture Devices Extend Medicine's Reach

An ISS experiment led to the development of medical ultrasound diagnostic techniques for long-distance use. Technology created to capture and transmit these ultrasound results over the Internet allows patients from professional athletes to mountain climbers to receive medical attention as soon as needed (*Spinoff* 2009).

Resistance Systems Provide Healthy Workouts

Developed to help astronauts perform vital exercise during long stays on the ISS, stretching elastomer technology now serves as an effective source of resistance

for workout machines on Earth, replicating the feel and results—but not the unwieldy bulk—of free weights (*Spinoff* 2001).

Programmable Ovens Let You Start Dinner from the Web

Engineers who designed the ISS Electric Power System created "embedded Web technology" which allows users to control devices remotely, including a commercial oven. With both heating and cooling capabilities, this oven can refrigerate a prepared dish until the programmable cooking cycle begins, allowing dinner to be perfectly cooked when the user arrives home (*Spinoff* 2005).

Aeroponic Gardens Help Plants Grow Faster and Healthier

A soil-less plant-growth experiment that proved healthy plant growth without the use of pesticides has enabled the development of a commercial aeroponic system. The sterile environment allows plants to grow disease-free, with 98-percent less water, and no pesticides (*Spinoff* 2006).

Systems Make Automobile Testing More Accurate

Automobile safety testing improved when a NASA charge coupled device camera—originally designed to track bar codes on parts for the robotic assembly of the ISS—was combined with a newly developed synthetic mask skin covering for crash test dummies. As a combined system, the technologies provide more precise, repeatable predictions of laceration injuries sustained in automobile accidents (*Spinoff* 2002).

Studies into ways to keep astronaut's muscles toned while living in microgravity conditions led to improved resistance training devices (above) here on Earth. An oven (below) that allows users to remotely control cooking times owes its brains to software designed to allow astronauts to operate experiments from anywhere on the ISS.

Experiments into growing plants as food sources for long-duration space flight have resulted in new methods for gardening here on Earth, techniques that enable plants to develop strong, healthy roots, with a minimal amount of soil.

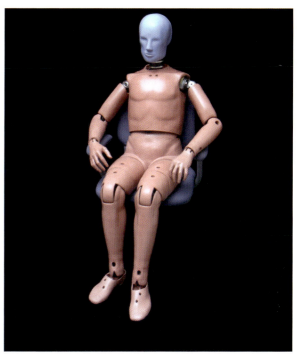

Components of a camera designed to aid robotic assembly of the ISS have been used to analyze injuries on crash test dummies, enabling engineers to design safer automobiles.

Air Purifiers Eliminate Pathogens, Preserve Food

NASA research into sustaining perishable foods for long-duration space missions resulted in the development of an air-cleaning device that eliminates airborne bacteria, mold, fungi, mycotoxins, viruses, volatile organic compounds, and odors (*Spinoff* 2009).

A robotic hand designed for conducting precision repairs on the ISS has found applications in surgical settings here on Earth.

Portable System Warns of Dangerous Pressure Changes

The Personal Cabin Pressure Altitude Monitor and Warning System is a hand-held, personal safety device to warn pilots of potentially dangerous or deteriorating cabin pressure altitude conditions before hypoxia becomes a threat. It was designed as a backup device for ISS crewmembers. Applications beyond aviation and aerospace include scuba diving, skydiving, mountain climbing, meteorology, altitude chambers, and underwater habitats (*Spinoff* 2003).

ISS Materials Research Leads to Improved Golf Clubs

A material designed for the space station aided in the development of Zeemet, a proprietary shape memory alloy for the golf industry. The Nicklaus Golf Company created a line of golf clubs using Zeemet inserts. Its super-elastic and high-damping attributes translate into more spin on the ball, greater control, and a solid feel (*Spinoff* 1997).

Robotics Offer Newfound Surgical Capabilities

Robotics designed for intricate repairs on the ISS find many industry uses, including a minimally invasive knee surgery procedure, where its precision control makes it ideal for inserting a very small implant (*Spinoff* 2008).

Life Support System Recycles Water

A water filtration system providing safe, affordable drinking water throughout the world is the result of work by NASA engineers who created the Regenerative Environmental Control and Life Support System, a complex system of devices intended to sustain the astronauts living on the ISS (*Spinoff* 2006).

The water recycling system designed for the ISS led to the development of a rugged, portable device that can bring clean water to remote areas of the planet.

LEDs Alleviate Pain, Speed Rehabilitation

Tiny light-emitting diode (LED) chips used to grow plants on the ISS are used for wound healing and chronic pain alleviation on Earth and have been successfully applied in cases of pediatric brain tumors and the prevention of oral mucositis in bone marrow transplant patients (*Spinoff* 2008).

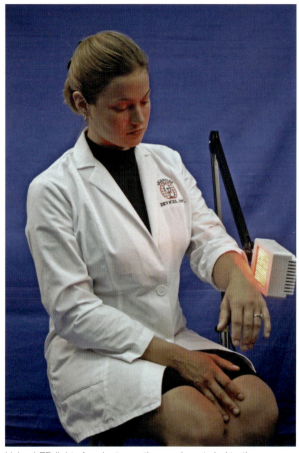

Using LED lights for plant growth experiments led to the development of a device proven to aid in the treatment of injuries.

Studies into astronaut exercise in space led to the development of a rehabilitation device that applies air pressure to a patient's lower body in order to unload weight, which reduces the stress placed on the lower body during rehabilitation.

'Anti-Gravity' Treadmills Speed Rehabilitation

Research into the biomechanics of exercise, using differential air pressure in space to mimic the Earth's gravity to prevent bone loss and muscle deterioration, led to the development of a treadmill that is now aiding patients with various neurological or musculoskeletal conditions (*Spinoff* 2009).

Food Supplement Reduces Fat, Improves Flavor

Extending the shelf life of food while still preserving flavor—key factors for long-duration space flight—led to the development of a fat substitute intended for use as a partial replacement for animal fat in beef patties and other normally high-fat meat products. The substitute can also be used in soups, sauces, bakery items, and desserts (*Spinoff* 2007).

Head-Mounted System Aids Vision

The Low Vision Enhancement System is a video headset that offers people with low vision a view of their surroundings equivalent to the image on a 5-foot television screen 4 feet from the viewer. For many people with low vision, it eases everyday activities such as reading, watching TV, and shopping. Researchers used NASA technology for computer processing of satellite images and head-mounted vision enhancement systems originally intended for the space station (*Spinoff* 1995). ❖

Executive Summary

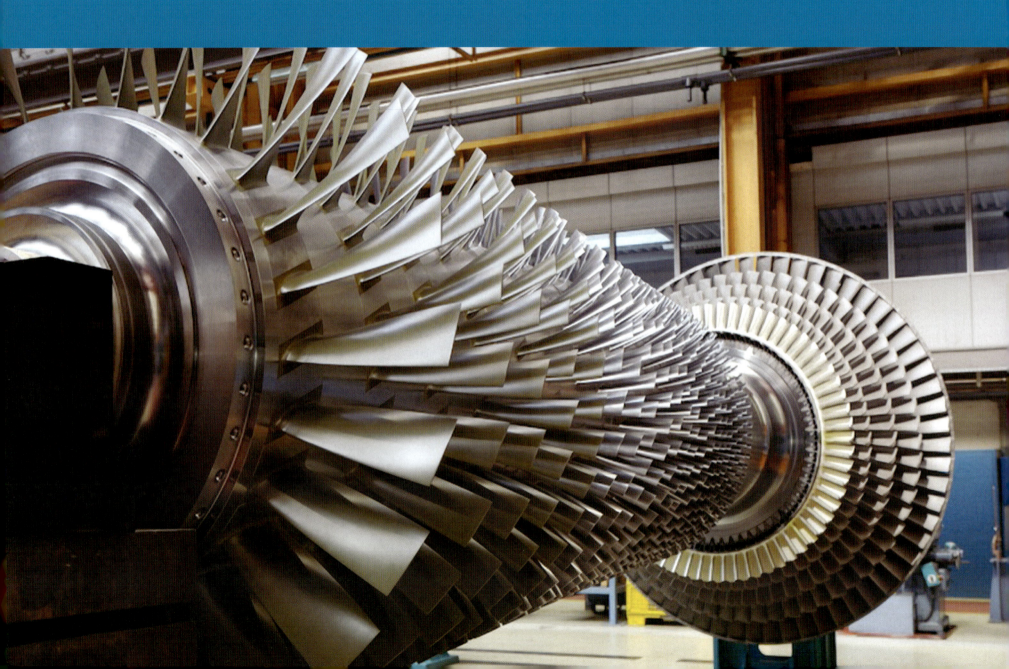

In accordance with congressional mandates cited in the National Aeronautics and Space Act of 1958 and the Technology Utilization Act of 1962, NASA has nurtured partnerships with the private sector to facilitate the transfer of Agency-developed technologies for the greater good of the public. These partnerships fuel economic and technological development nationally and globally, resulting in commercial products and services enabled on Earth by NASA's missions to the stars. Since 1976, NASA *Spinoff* has profiled the most compelling of these technologies, annually highlighting the best and brightest of partnerships and innovations in the fields of health and medicine, transportation, public safety, consumer goods, environmental resources, computer technology, and industrial productivity.

Executive Summary

NASA *Spinoff* highlights the Agency's most significant research and development activities and the successful transfer of NASA technology, showcasing the cutting-edge research being done by the Nation's top technologists and the practical benefits that come back down to Earth in the form of tangible products that make our lives better. The benefits featured in this year's issue include:

Health and Medicine

Burnishing Techniques Strengthen Hip Implants

In the late 1990s, Lambda Research Inc., of Cincinnati, Ohio, received Small Business Innovation Research (SBIR) awards from Glenn Research Center to demonstrate low plasticity burnishing (LPB) on metal engine components. By producing a thermally stable deep layer of compressive residual stress, LPB significantly strengthened turbine alloys. After Lambda patented the process, the Federal Aviation Administration accepted LPB for repair and alteration of commercial aircraft components, the U.S. Department of Energy found LPB suitable for treating nuclear waste containers at Yucca Mountain. Data from the U.S. Food and Drug Administration confirmed LPB to completely eliminate the occurrence of fretting fatigue failures in modular hip implants.
page 38

Signal Processing Methods Monitor Cranial Pressure

Dr. Norden Huang, of Goddard Space Flight Center, invented a set of algorithms (called the Hilbert-Huang Transform, or HHT) for analyzing nonlinear and nonstationary signals that developed into a user-friendly signal processing technology for analyzing time-varying processes. At an auction managed by Ocean Tomo Federal Services LLC, licenses of 10 U.S. patents and 1 domestic patent application related to HHT were sold to DynaDx Corporation, of Mountain View, California. DynaDx is now using the licensed NASA technology for medical diagnosis and prediction of brain blood flow-related problems, such as stroke, dementia, and traumatic brain injury.
page 40

Ultraviolet-Blocking Lenses Protect, Enhance Vision

To combat the harmful properties of light in space, as well as that of artificial radiation produced during laser and welding work, Jet Propulsion Laboratory (JPL) scientists developed a lens capable of absorbing, filtering, and scattering the dangerous light while not obstructing vision. SunTiger Inc.—now Eagle Eyes Optics, of Calabasas, California—was formed to market a full line of sunglasses based on the JPL discovery that promised 100-percent elimination of harmful wavelengths and enhanced visual clarity. The technology was recently inducted into the Space Technology Hall of Fame.
page 42

Hyperspectral Systems Increase Imaging Capabilities

In 1983, NASA started developing hyperspectral systems to image in the ultraviolet and infrared wavelengths. In 2001, the first on-orbit hyperspectral imager, Hyperion, was launched aboard the Earth Observing-1 spacecraft. Based on the hyperspectral imaging sensors used in Earth observation satellites, Stennis Space Center engineers and Institute for Technology Development researchers collaborated on a new design that was smaller and used an improved scanner. Featured in *Spinoff* 2007, the technology is now exclusively licensed by Themis Vision Systems LLC, of Richmond, Virginia, and is widely used in medical and life sciences, defense and security, forensics, and microscopy.
page 44

Transportation

Programs Model the Future of Air Traffic Management

Through Small Business Innovation Research (SBIR) contracts with Ames Research Center, Intelligent Automation Inc., based in Rockville, Maryland, advanced specialized software the company had begun developing with U.S. Department of Defense funding. The agent-based infrastructure now allows NASA's Airspace Concept Evaluation System to explore ways of improving the utilization of the National Airspace System (NAS), providing flexible modeling of every part of the NAS down to individual planes, airports, control centers, and even weather. The software has been licensed to a number of aerospace and robotics customers, and has even been used to model the behavior of crowds.
page 48

Tail Rotor Airfoils Stabilize Helicopters, Reduce Noise

Founded by former Ames Research Center engineer Jim Van Horn, Van Horn Aviation of Tempe, Arizona, built upon a Langley Research Center airfoil design to create a high performance aftermarket tail rotor for the popular Bell 206 helicopter. The highly durable rotor has a lifetime twice that of the original equipment manufacturer blade, reduces noise by 40 percent, and displays enhanced performance at high altitudes. These improvements benefit helicopter performance for law enforcement, military training, wildfire and pipeline patrols, and emergency medical services.
page 50

Personal Aircraft Point to the Future of Transportation
NASA's Small Business Innovation Research (SBIR) and Small Business Technology Transfer (STTR) programs, as well as a number of Agency innovations, have helped Duluth, Minnesota-based Cirrus Design Corporation become one of the world's leading manufacturers of general aviation aircraft. SBIRs with Langley Research Center provided the company with cost-effective composite airframe manufacturing methods, while crashworthiness testing at the Center increased the safety of its airplanes. Other NASA-derived technologies on Cirrus SR20 and SR22 aircraft include synthetic vision systems that help pilots navigate and full-plane parachutes that have saved the lives of more than 30 Cirrus pilots and passengers to date. Today, the SR22 is the world's top-selling Federal Aviation Administration (FAA)-certified single-engine airplane.
page 52

Ducted Fan Designs Lead to Potential New Vehicles
In 1994, aerospace engineers Rob Bulaga and Mike Moshier formed Trek Aerospace Inc., based in Folsom, California, to develop personal air vehicles using a novel ducted fan design. The company relied on Ames Research Center for a great deal of testing, the results of which have provided greater lift, lowered weight, more power, and improved maneuverability. The technology has been applied to three models: the Dragonfly UMR-1, the Springtail EFV, and the OVIWUN, a small-scale version that is for sale through the company's Web site. It is safer than a manned vehicle, and its size makes it relatively difficult for it to damage itself during test flights the way a larger mass, faster craft could.
page 56

Winglets Save Billions of Dollars in Fuel Costs
The upturned ends now featured on many airplane wings are saving airlines billions of dollars in fuel costs. Called winglets, the drag-reducing technology was advanced through the research of Langley Research Center engineer Richard Whitcomb and through flight tests conducted at Dryden Flight Research Center. Seattle-based Aviation Partners Boeing—a partnership between Aviation Partners Inc., of Seattle, and The Boeing Company, of Chicago—manufactures Blended Winglets, a unique design featured on Boeing aircraft around the world. These winglets have saved more than 2 billion gallons of jet fuel to date, representing a cost-savings of more than $4 billion and a reduction of almost 21.5 million tons in carbon dioxide emissions.
page 58

Sensor Systems Collect Critical Aerodynamics Data
With the support of Small Business Innovation Research (SBIR) contracts with Dryden Flight Research Center, Tao of Systems Integration Inc. developed sensors and other components that will ultimately form a first-of-its-kind, closed-loop system for detecting, measuring, and controlling aerodynamic forces and moments in flight. The Hampton, Virginia-based company commercialized three of the four planned components, which provide sensing solutions for customers such as Boeing, General Electric, and BMW and are used for applications such as improving wind turbine operation and optimizing air flow from air conditioning systems. The completed system may one day enable flexible-wing aircraft with flight capabilities like those of birds.
page 60

Coatings Extend Life of Engines and Infrastructure
MesoCoat Inc., of Euclid, Ohio, collaborated with Glenn Research Center to provide thermal barrier coating (TBC) technology, developed by Glenn researcher Dongming Zhu, to enhance the lifespan and performance of engines in U.S. Air Force legacy aircraft. The TBC reduces thermal stresses on engine parts, increasing component life by 50 percent. MesoCoat is also producing metal cladding technology that may soon provide similar life-lengthening benefits for the Nation's infrastructure. Through a Space Act Agreement with Glenn, the company employs the Center's high-density infrared arc lamp system to bond its cladding materials for demonstration prototypes; the coating technology can prevent corrosion on metal beams, pipes, and rebar for up to 100 years.
page 62

Public Safety

Radiometers Optimize Local Weather Prediction
Radiometrics Corporation, headquartered in Boulder, Colorado, engaged in Small Business Innovation Research (SBIR) agreements with Glenn Research Center that resulted in a pencil-beam radiometer designed to detect supercooled liquid along flight paths—a prime indicator of dangerous icing conditions. The company has brought to market a modular radiometer that resulted from the SBIR work. Radiometrics' radiometers are used around the world as key tools for detecting icing conditions near airports and for the prediction of weather conditions like fog and convective storms, which are known to produce hail, strong winds, flash floods, and tornadoes. They are also employed for oceanographic research and soil moisture studies.
page 66

Energy-Efficient Systems Eliminate Icing Danger for UAVs

Ames Research Center engineer Leonard Haslim invented an anti-icing technology called an electroexpulsive separation system, which uses mechanical force to shatter potentially dangerous ice buildup on an aircraft surface. Temecula, California-based Ice Management Systems (now known as IMS-ESS) licensed the technology from Ames and has discovered a niche market for the lightweight, energy-efficient technology: unmanned aerial vehicles (UAVs). IMS-ESS systems now prevent damaging ice accumulation on military UAVs, allowing the vehicles to carry out crucial missions year round.
page 68

Rocket-Powered Parachutes Rescue Entire Planes

Small Business Innovation Research (SBIR) contracts with Langley Research Center helped BRS Aerospace, of Saint Paul, Minnesota, to develop technology that has saved 246 lives to date. The company's whole aircraft parachute systems deploy in less than 1 second thanks to solid rocket motors and are capable of arresting the descent of a small aircraft, lowering it safely to the ground. BRS has sold more than 30,000 systems worldwide, and the technology is now standard equipment on many of the world's top-selling aircraft. Parachutes for larger airplanes are in the works.
page 70

Technologies Advance UAVs for Science, Military

A Space Act Agreement with Goddard Space Flight Center and West Virginia University enabled Aurora Flight Sciences Corporation, of Manassas, Virginia, to develop cost-effective composite manufacturing capabilities and open a facility in West Virginia. The company now employs 160 workers at the plant, tasked with crafting airframe components for the Global Hawk unmanned aerial vehicle (UAV) program. While one third of the company's workforce focuses on Global Hawk production, the rest of the company develops advanced UAV technologies that are redefining traditional approaches to unmanned aviation. Since the company's founding, Aurora's cutting-edge work has been supported with funding from NASA's Small Business Innovation Research (SBIR) and Small Business Technology Transfer (STTR) programs.
page 72

Inflatable Antennas Support Emergency Communication

Glenn Research Center awarded Small Business Innovation Research (SBIR) contracts to ManTech SRS Technologies, of Newport Beach, California, to develop thin film inflatable antennas for space communication. With additional funding, SRS modified the concepts for ground-based inflatable antennas. GATR (Ground Antenna Transmit and Receive) Technologies, of Huntsville, Alabama, licensed the technology and refined it to become the world's first inflatable antenna certified by the Federal Communications Commission. Capable of providing Internet access, voice over Internet protocol, e-mail, video teleconferencing, broadcast television, and other high-bandwidth communications, the systems have provided communication during the wildfires in California, after Hurricane Katrina in Mississippi, and following the 2010 Haiti earthquake.
page 74

Smart Sensors Assess Structural Health

NASA frequently inspects launch vehicles, fuel tanks, and other components for structural damage. To perform quick evaluation and monitoring, the Agency pursues the development of structural health monitoring systems. In 2001, Acellent Technologies Inc., of Sunnyvale, California, received Small Business Innovation Research (SBIR) funding from Marshall Space Flight Center to develop a hybrid Stanford Multi-Actuator Receiver Transduction (SMART) Layer for aerospace vehicles and structures. As a result, Acellent expanded the technology's capability and now sells it to aerospace and automotive companies; construction, energy, and utility companies; and the defense, space, transportation, and energy industries for structural condition monitoring, damage detection, crack growth monitoring, and other applications.
page 76

Hand-Held Devices Detect Explosives and Chemical Agents

Ion Applications Inc., of West Palm Beach, Florida, partnered with Ames Research Center through Small Business Innovation Research (SBIR) agreements to develop a miniature version ion mobility spectrometer (IMS). While NASA was interested in the instrument for detecting chemicals during exploration of distant planets, moons, and comets, the company has incorporated the technology into a commercial hand-held IMS device for use by the military and other public safety organizations. Capable of detecting and identifying molecules with part-per-billion sensitivity, the technology now provides soldiers with portable explosives and chemical warfare agent detection. The device is also being adapted for detecting drugs and is employed in industrial processes such as semiconductor manufacturing.
page 78

Terahertz Tools Advance Imaging for Security, Industry

Picometrix, a wholly owned subsidiary of Advanced Photonix Inc. (API), of Ann Arbor, Michigan, invented the world's first commercial terahertz system. The company improved the portability and capabilities of their systems through Small Business Innovation Research (SBIR) agreements with Langley Research Center to provide terahertz imaging capabilities for inspecting the space shuttle external tanks and orbiters. Now API's systems make use of the unique imaging capacity of terahertz radiation on manufacturing floors, for thickness measurements of coatings, pharmaceutical tablet production, and even art conservation.
page 80

Consumer Goods

LED Systems Target Plant Growth

To help develop technologies for growing edible biomass (food crops) in space, Kennedy Space Center partnered with Orbital Technologies Corporation (ORBITEC), of Madison, Wisconsin, through the Small Business Innovation Research (SBIR) program. One result of this research was the High Efficiency Lighting with Integrated Adaptive Control (HELIAC) system, components of which have been incorporated into a variety of agricultural greenhouse and consumer aquarium lighting features. The new lighting systems can be adapted to a specific plant species during a specific growth stage, allowing maximum efficiency in light absorption by all available photosynthetic tissues.
page 84

Aerogels Insulate Against Extreme Temperatures

In 1992, NASA started to pursue the development of aerogel for cryogenic insulation. Kennedy Space Center awarded Small Business Innovation Research (SBIR) contracts to Aspen Systems Inc., of Marlborough, Massachusetts, that resulted in a new manufacturing process and a new flexible, durable, easy-to-use form of aerogel. Aspen Systems formed Aspen Aerogels Inc., in Northborough, Massachusetts, to market the product, and by 2009, the company had become the leading provider of aerogel in the United States, producing nearly 20 million square feet per year. With an array of commercial applications, the NASA-derived aerogel has most recently been applied to protect and insulate people's hands and feet.
page 86

Image Sensors Enhance Camera Technologies

In the 1990s, a Jet Propulsion Laboratory team led by Eric Fossum researched ways of improving complementary metal-oxide semiconductor (CMOS) image sensors in order to miniaturize cameras on spacecraft while maintaining scientific image quality. Fossum's team founded a company to commercialize the resulting CMOS active pixel sensor. Now called the Aptina Imaging Corporation, based in San Jose, California, the company has shipped over 1 billion sensors for use in applications such as digital cameras, camera phones, Web cameras, and automotive cameras. Today, one of every three cell phone cameras on the planet feature Aptina's sensor technology.
page 90

Lightweight Material Patches Allow for Quick Repairs

Cornerstone Research Group Inc., of Dayton, Ohio, has been the recipient of 16 Small Business Innovation Research (SBIR) contracts with NASA with a variety of different focuses, including projects like creating inflatable structures for radio frequency antennas and, most recently, healable polymer matrix composites for future space vehicles. One of its earlier SBIR contracts, with Kennedy Space Center, led to the development of a new type of structural patch for a variety of consumer uses: Rubbn'Repair, for automotive uses; and Rec'Repair for the outdoors and adventure market. Both are flexible, heat-activated structural patches.
page 92

Nanomaterials Transform Hairstyling Tools

Dr. Dennis Morrison, a former scientist at Johnson Space Center, conducted research on microcapsules that were developed in space and designed to deliver drugs to cancerous tumors. This work led to research on nanoceramic materials, and in 2001, Morrison shared his expertise with Farouk Shami, the owner of Farouk Systems Inc., of Houston, Texas. After learning more, Shami developed a ceramic composite for his CHI (Cationic Hydration Interlink) hairstyling irons, brushes, nail lacquers, and hair dryers. Morrison also used his NASA research expertise as a platform to incorporate nanosilver and near-infrared light into the products.
page 94

Do-It-Yourself Additives Recharge Auto Air Conditioning

In planning for a return mission to the Moon, NASA aimed to improve the thermal control systems that keep astronauts comfortable and cool while inside a spacecraft. Goddard Space Flight Center awarded a Small Business Innovation Research (SBIR) contract to Mainstream Engineering Corporation, of Rockledge, Florida, to develop a chemical/mechanical heat pump. While working on the design, Mainstream Engineering came up with a unique liquid additive called QwikBoost to enhance the performance of the advanced heat pump design.

IDQ Inc., of Garland, Texas, exclusively licensed the technology and incorporates it into its line of Arctic Freeze products for automotive air conditioning applications.
page 96

Environmental Resources

Systems Analyze Water Quality in Real Time

A water analyzer developed under Small Business Innovation Research (SBIR) contracts with Kennedy Space Center now monitors treatment processes at water and wastewater facilities around the world. Originally designed to provide real-time detection of nutrient levels in hydroponic solutions for growing plants in space, the ChemScan analyzer, produced by ASA Analytics Inc., of Waukesha, Wisconsin, utilizes spectrometry and chemometric algorithms to automatically analyze multiple parameters in the water treatment process with little need for maintenance, calibration, or operator intervention. The company has experienced a compound annual growth rate of 40 percent over its 15-year history as a direct result of the technology's success.
page 100

Compact Radiometers Expand Climate Knowledge

To gain a better understanding of Earth's water, energy, and carbon cycles, NASA plans to embark on the Soil Moisture Active and Passive mission in 2015. To prepare, Goddard Space Flight Center provided Small Business Innovation Research (SBIR) funding to ProSensing Inc., of Amherst, Massachusetts, to develop a compact ultrastable radiometer for sea surface salinity and soil moisture mapping. ProSensing incorporated small, low-cost, high-performance elements into just a few circuit boards and now offers two lightweight radiometers commercially. Government research agencies, university research groups, and large corporations around the world are using the devices for mapping soil moisture, ocean salinity, and wind speed.
page 102

Energy Servers Deliver Clean, Affordable Power

K.R. Sridhar developed a fuel cell device for Ames Research Center, that could use solar power to split water into oxygen for breathing and hydrogen for fuel on Mars. Sridhar saw the potential of the technology, when reversed, to create clean energy on Earth. He founded Bloom Energy, of Sunnyvale, California, to advance the technology. Today, the Bloom Energy Server is providing cost-effective, environmentally friendly energy to a host of companies such as eBay, Google, and The Coca-Cola Company. Bloom's NASA-derived Energy Servers generate energy that is about 67-percent cleaner than a typical coal-fired power plant when using fossil fuels and 100-percent cleaner with renewable fuels.
page 104

Solutions Remediate Contaminated Groundwater

During the Apollo Program, NASA workers used chlorinated solvents to clean rocket engine components at launch sites. These solvents, known as dense non-aqueous phase liquids, had contaminated launch facilities to the point of near-irreparability. Dr. Jacqueline Quinn and Dr. Kathleen Brooks Loftin of Kennedy Space Center partnered with researchers from the University of Central Florida's chemistry and engineering programs to develop technology capable of remediating the area without great cost or further environmental damage. They called the new invention Emulsified Zero-Valent Iron (EZVI). The groundwater remediation compound is cleaning up polluted areas all around the world and is, to date, NASA's most-licensed technology.
page 106

Bacteria Provide Cleanup of Oil Spills, Wastewater

Through Small Business Innovation Research (SBIR) contracts with Marshall Space Flight Center, Micro-Bac International Inc., of Round Rock, Texas, developed a phototrophic cell for water purification in space. Inside the cell: millions of photosynthetic bacteria. Micro-Bac proceeded to commercialize the bacterial formulation it developed for the SBIR project. The formulation is now used for the remediation of wastewater systems and waste from livestock farms and food manufacturers. Strains of the SBIR-derived bacteria also feature in microbial solutions that treat environmentally damaging oil spills, such as that resulting from the catastrophic 2010 Deepwater Horizon oil rig explosion in the Gulf of Mexico.
page 108

Reflective Coatings Protect People and Animals

Led by Marshall Space Flight Center, NASA engineers called upon National Metalizing of Cranbury, New Jersey, to help create a reflective sunshield to deploy on Skylab in place of a shield that was lost during launch in 1973. Years later, a former employee for National Metalizing founded Advanced Flexible Materials (AFM) Inc., of Petaluma, California, and utilized the radiant barrier technology in the public domain to produce a variety of products such as wraps to keep marathon finishers safe from hypothermia as well as a lining for mittens and vests. Recently, the material helped to keep manatees warm as they were lifted from the water as part of a tag-and-release program.
page 110

Computer Technology

Innovative Techniques Simplify Vibration Analysis

In the early years of development, Marshall Space Flight Center engineers encountered challenges related to components in the space shuttle main engine. To assess the problems, they evaluated the effects of vibration and oscillation. To enhance the method of vibration signal analysis, Marshall awarded Small Business Innovation Research (SBIR) contracts to AI Signal Research, Inc. (ASRI), in Huntsville, Alabama. ASRI developed a software package called PC-SIGNAL that NASA now employs on a daily basis, and in 2009, the PKP-Module won Marshall's "Software of the Year" award. The technology is also used in many industries: aircraft and helicopter, rocket engine manufacturing, transportation, and nuclear power.
page 114

Modeling Tools Predict Flow in Fluid Dynamics

Because rocket engines operate under extreme temperature and pressure, they present a unique challenge to designers who must test and simulate the technology. To this end, CRAFT Tech Inc., of Pipersville, Pennsylvania, won Small Business Innovation Research (SBIR) contracts from Marshall Space Flight Center to develop software to simulate cryogenic fluid flows and related phenomena. CRAFT Tech enhanced its CRUNCH CFD (computational fluid dynamics) software to simulate phenomena in various liquid propulsion components and systems. Today, both government and industry clients in the aerospace, utilities, and petrochemical industries use the software for analyzing existing systems as well as designing new ones.
page 116

Verification Tools Secure Online Shopping, Banking

Just like rover or rocket technology sent into space, the software that controls these technologies must be extensively tested to ensure reliability and effectiveness. Ames Research Center invented the open-source Java Pathfinder (JPF) toolset for the deep testing of Java-based programs. Fujitsu Labs of America Inc., based in Sunnyvale, California, improved the capabilities of the JPF Symbolic Pathfinder tool, establishing the tool as a means of thoroughly testing the functionality and security of Web-based Java applications such as those used for Internet shopping and banking.
page 118

Toolsets Maintain Health of Complex Systems

First featured in *Spinoff* 2001, Qualtech Systems Inc. (QSI), of Wethersfield, Connecticut, adapted its Testability, Engineering, and Maintenance System (TEAMS) toolset under Small Business Innovation Research (SBIR) contracts from Ames Research Center to strengthen NASA's systems health management approach for its large, complex, and interconnected systems. Today, six NASA field centers utilize the TEAMS toolset, including TEAMS-Designer, TEAMS-RT, TEAMATE, and TEAMS-RDS. TEAMS is also being used on industrial systems that generate power, carry data, refine chemicals, perform medical functions, and produce semiconductor wafers. QSI finds TEAMS can lower costs by decreasing problems requiring service by 30 to 50 percent.
page 120

Framework Resources Multiply Computing Power

As an early proponent of grid computing, Ames Research Center awarded Small Business Innovation Research (SBIR) funding to 3DGeo Development Inc., of Santa Clara, California, (now FusionGeo Inc., of The Woodlands, Texas) to demonstrate a virtual computer environment that linked geographically dispersed computer systems over the Internet to help solve large computational problems. By adding to an existing product, FusionGeo enabled access to resources for calculation- or data-intensive applications whenever and wherever they were needed. Commercially available as Accelerated Imaging and Modeling, the product is used by oil companies and seismic service companies, which require large processing and data storage capacities.
page 122

Tools Automate Spacecraft Testing, Operation

NASA began the Small Explorer (SMEX) program to develop spacecraft to advance astrophysics and space physics. As one of the entities supporting software development at Goddard Space Flight Center, the Hammers Company Inc. (tHC Inc.), of Greenbelt, Maryland, developed the Integrated Test and Operations System to support SMEX. Later, the company received additional Small Business Innovation Research (SBIR) funding from Goddard for a tool to facilitate the development of flight software called VirtualSat. NASA uses the tools to support 15 satellites, and the aerospace industry is using them to develop science instruments, spacecraft computer systems, and navigation and control software.
page 124

GPS Software Packages Deliver Positioning Solutions

To determine a spacecraft's position, the Jet Propulsion Laboratory (JPL) developed an innovative software program called the GPS (global positioning system)-Inferred Positioning System and Orbit Analysis Simulation Software, abbreviated as GIPSY-OASIS, and also developed

Real-Time GIPSY (RTG) for certain time-critical applications. First featured in *Spinoff* 1999, JPL has released hundreds of licenses for GIPSY and RTG, including to Longmont, Colorado-based DigitalGlobe. Using the technology, DigitalGlobe produces satellite imagery with highly precise latitude and longitude coordinates and then supplies it for uses within defense and intelligence, civil agencies, mapping and analysis, environmental monitoring, oil and gas exploration, infrastructure management, Internet portals, and navigation technology.
page 126

Solid-State Recorders Enhance Scientific Data Collection

Under Small Business Innovation Research (SBIR) contracts with Goddard Space Flight Center, SEAKR Engineering Inc., of Centennial, Colorado, crafted a solid-state recorder (SSR) to replace the tape recorder onboard a Spartan satellite carrying NASA's Inflatable Antenna Experiment. Work for that mission and others has helped SEAKR become the world leader in SSR technology for spacecraft. The company has delivered more than 100 systems, more than 85 of which have launched onboard NASA, military, and commercial spacecraft—including imaging satellites that provide much of the high-resolution imagery for online mapping services like Google Earth.
page 128

Computer Models Simulate Fine Particle Dispersion

Through a NASA Seed Fund partnership with DEM Solutions Inc., of Lebanon, New Hampshire, scientists at Kennedy Space Center refined existing software to study the electrostatic phenomena of granular and bulk materials as they apply to planetary surfaces. The software, EDEM, allows users to import particles and obtain accurate representations of their shapes for modeling purposes, such as simulating bulk solids behavior, and was enhanced to be able to more accurately model fine, abrasive, cohesive particles. These new EDEM capabilities can be applied in many industries unrelated to space exploration and have been adopted by several prominent U.S. companies, including John Deere, Pfizer, and Procter & Gamble.
page 130

Industrial Productivity

Composite Sandwich Technologies Lighten Components

Leveraging its private resources with several Small Business Innovation Research (SBIR) contracts with both NASA and the U.S. Department of Defense, WebCore Technologies LLC, of Miamisburg, Ohio, developed a fiber-reinforced foam sandwich panel it calls TYCOR that can be used for a wide variety of industrial and consumer applications. Testing at Glenn Research Center's Ballistic Impact Facility demonstrated that the technology was able to exhibit excellent damage localization and stiffness during impact. The patented and trademarked material has found use in many demanding applications, including marine, ground transportation, mobile shelters, bridges, and most notably, wind turbines.
page 134

Cameras Reveal Elements in the Short Wave Infrared

Goodrich ISR Systems Inc. (formerly Sensors Unlimited Inc.), based out of Princeton, New Jersey, received Small Business Innovation Research (SBIR) contracts from the Jet Propulsion Laboratory, Marshall Space Flight Center, Kennedy Space Center, Goddard Space Flight Center, Ames Research Center, Stennis Space Center, and Langley Research Center to assist in advancing and refining indium gallium arsenide imaging technology. Used on the Lunar Crater Observation and Sensing Satellite (LCROSS) mission in 2009 for imaging the short wave infrared wavelengths, the technology has dozens of applications in military, security and surveillance, machine vision, medical, spectroscopy, semiconductor inspection, instrumentation, thermography, and telecommunications.
page 136

Deformable Mirrors Correct Optical Distortions

By combining the high sensitivity of space telescopes with revolutionary imaging technologies consisting primarily of adaptive optics, the Terrestrial Planet Finder is slated to have imaging power 100 times greater than the Hubble Space Telescope. To this end, Boston Micromachines Corporation, of Cambridge, Massachusetts, received Small Business Innovation Research (SBIR) contracts from the Jet Propulsion Laboratory for space-based adaptive optical technology. The work resulted in a microelectromechanical systems (MEMS) deformable mirror (DM) called the Kilo-DM. The company now offers a full line of MEMS DMs, which are being used in observatories across the world, in laser communication, and microscopy.
page 138

Stitching Techniques Advance Optics Manufacturing

Because NASA depends on the fabrication and testing of large, high-quality aspheric (non-spherical) optics for applications like the James Webb Space Telescope, it sought an improved method for measuring large aspheres. Through Small Business Innovation Research (SBIR) awards from Goddard Space Flight Center, QED Technologies, of Rochester, New York, upgraded and enhanced its stitching technology for aspheres. QED developed the SSI-A, which earned the company an "R&D 100" award, and also developed a breakthrough machine

tool called the aspheric stitching interferometer. The equipment is applied to advanced optics in telescopes, microscopes, cameras, medical scopes, binoculars, and photolithography.
page 140

Compact, Robust Chips Integrate Optical Functions
Located in Bozeman, Montana, AdvR Inc. has been an active partner in NASA's Small Business Innovation Research (SBIR) and Small Business Technology Transfer (STTR) programs. Langley Research Center engineers partnered with AdvR through the SBIR program to develop new, compact, lightweight electro-optic components for remote sensing systems. While the primary customer for this technology will be NASA, AdvR foresees additional uses for its NASA-derived circuit chip in the fields of academic and industrial research—anywhere that compact, low-cost, stabilized single-frequency lasers are needed.
page 142

Fuel Cell Stations Automate Processes, Catalyst Testing
Glenn Research Center looks for ways to improve fuel cells, which are an important source of power for space missions, as well as the equipment used to test fuel cells. With Small Business Innovation Research (SBIR) awards from Glenn, Lynntech Inc., of College Station, Texas, addressed a major limitation of fuel cell testing equipment. Five years later, the company obtained a patent and provided the equipment to the commercial world. Now offered through TesSol Inc., of Battle Ground, Washington, the technology is used for fuel cell work, catalyst testing, sensor testing, gas blending, and other applications. It can be found at universities, national laboratories, and businesses around the world.
page 144

Onboard Systems Record Unique Videos of Space Missions
Ecliptic Enterprises Corporation, headquartered in Pasadena, California, provided onboard video systems for rocket and space shuttle launches before it was tasked by Ames Research Center to craft the Data Handling Unit that would control sensor instruments onboard the Lunar Crater Observation and Sensing Satellite (LCROSS) spacecraft. The technological capabilities the company acquired on this project, as well as those gained developing a high-speed video system for monitoring the parachute deployments for the Orion Pad Abort Test Program at Dryden Flight Research Center, have enabled the company to offer high-speed and high-definition video for geosynchronous satellites and commercial space missions, providing remarkable footage that both informs engineers and inspires the imagination of the general public.
page 146

Space Research Results Purify Semiconductor Materials
One of NASA's Commercial Space Centers, the Space Vacuum Epitaxy Center (SVEC), had a mission to create thin film semiconductor materials and devices through the use of vacuum growth technologies. In partnership with Johnson Space Center, researchers spent years in the lab where they advanced a technique called molecular beam epitaxy. In 1997, researchers from the SVEC formed a company called Applied Optoelectronics Inc., of Sugar Land, Texas, to fabricate devices using the advanced techniques and knowledge. Today, the company develops and manufactures optical devices for fiber optic networks including cable television, wireless, telecommunications, data communications, and fiber-to-the-home applications.
page 148

Toolkits Control Motion of Complex Robotics
Control and simulation software developed under Small Business Innovation Research (SBIR) contracts with Johnson Space Center is now providing user-friendly, optimized design and control of innovative robots used for military, agriculture, health care, and industrial applications. Created by Energid Technologies Corporation, of Cambridge, Massachusetts, the Actin toolkit provides for fluid robot motion, enhancing strength and accuracy while avoiding collisions and joint limits. Actin provides control capabilities for virtually any kind of robot, any joint type or tool type, and for any number of joints, degrees of freedom, and branches. In addition, the software provides powerful simulation capabilities, allowing developers to rapidly devise and test robot designs before the robot is built.
page 150

NASA Technologies Enhance Our Lives on Earth

International Space Station

Space Telescopes and Deep Space Exploration

Satellites and Imaging Technology

Innovative technologies from NASA's space and aeronautics missions (above) transfer as benefits to many sectors of society (below). Each benefit featured in *Spinoff* 2010 is listed with an icon that corresponds to the mission from which the technology originated. These NASA-derived technologies, when transferred to the public sector:

Health and Medicine

 Strengthen Hip Implants
page 38

 Monitor Cranial Pressure
page 40

 Protect, Enhance Vision
page 42

 Increase Imaging Capabilities
page 44

Transportation

 Model the Future of Air Traffic Management
page 48

 Stabilize Helicopters, Reduce Noise
page 50

 Point to the Future of Transportation
page 52

 Lead to Potential New Vehicles
page 56

 Save Billions of Dollars in Fuel Costs
page 58

 Collect Critical Aerodynamics Data
page 60

 Extend Life of Engines and Infrastructure
page 62

Public Safety

 Optimize Local Weather Prediction
page 66

 Eliminate Icing Danger for UAVs
page 68

 Rescue Entire Planes
page 70

 Advance UAVs for Science, Military
page 72

 Support Emergency Communication
page 74

 Assess Structural Health
page 76

 Detect Explosives and Chemical Agents
page 78

 Advance Imaging for Security, Industry
page 80

Consumer Goods

 Target Plant Growth
page 84

 Insulate Against Extreme Temperatures
page 86

 Enhance Camera Technologies
page 90

 Allow for Quick Repairs
page 92

 Transform Hairstyling Tools
page 94

 Recharge Auto Air Conditioning
page 96

Space Transportation

Environmental Resources

 Analyze Water Quality in Real Time
page 100

 Expand Climate Knowledge
page 102

 Deliver Clean, Affordable Power
page 104

 Remediate Contaminated Groundwater
page 106

 Provide Cleanup of Oil Spills, Wastewater
page 108

 Protect People and Animals
page 110

Astronaut Life Support

Computer Technology

 Simplify Vibration Analysis
page 114

 Predict Flow in Fluid Dynamics
page 116

 Secure Online Shopping, Banking
page 118

 Maintain Health of Complex Systems
page 120

 Multiply Computing Power
page 122

 Automate Spacecraft Testing, Operation
page 124

 Deliver Positioning Solutions
page 126

 Enhance Scientific Data Collection
page 128

 Simulate Fine Particle Dispersion
page 130

Aeronautics Research

Industrial Productivity

 Lighten Components
page 134

 Reveal Elements in the Short Wave Infrared
page 136

 Correct Optical Distortions
page 138

 Advance Optics Manufacturing
page 140

 Integrate Optical Functions
page 142

 Automate Processes, Catalyst Testing
page 144

 Record Unique Videos of Space Missions
page 146

 Purify Semiconductor Materials
page 148

 Control Motion of Complex Robotics
page 150

NASA Partnerships Across the Nation

Health and Medicine

1. Burnishing Techniques Strengthen Hip Implants (OH)
2. Signal Processing Methods Monitor Cranial Pressure (CA)
3. Ultraviolet-Blocking Lenses Protect, Enhance Vision (CA)
4. Hyperspectral Systems Increase Imaging Capabilities (VA)

Transportation

5. Programs Model the Future of Air Traffic Management (MD)
6. Tail Rotor Airfoils Stabilize Helicopters, Reduce Noise (AZ)
7. Personal Aircraft Point to the Future of Transportation (MN)
8. Ducted Fan Designs Lead to Potential New Vehicles (CA)
9. Winglets Save Billions of Dollars in Fuel Costs (WA)
10. Sensor Systems Collect Critical Aerodynamics Data (VA)
11. Coatings Extend Life of Engines and Infrastructure (OH)

Public Safety

12. Radiometers Optimize Local Weather Prediction (CO)
13. Energy-Efficient Systems Eliminate Icing Danger for UAVs (CA)
14. Rocket-Powered Parachutes Rescue Entire Planes (MN)
15. Technologies Advance UAVs for Science, Military (VA)
16. Inflatable Antennas Support Emergency Communication (AL)
17. Smart Sensors Assess Structural Health (CA)
18. Hand-Held Devices Detect Explosives and Chemical Agents (FL)
19. Terahertz Tools Advance Imaging for Security, Industry (MI)

Consumer Goods

20. LED Systems Target Plant Growth (WI)
21. Aerogels Insulate Against Extreme Temperatures (MA)
22. Image Sensors Enhance Camera Technologies (CA)
23. Lightweight Material Patches Allow for Quick Repairs (OH)
24. Nanomaterials Transform Hairstyling Tools (TX)
25. Do-It-Yourself Additives Recharge Auto Air Conditioning (TX)

Environmental Resources

26. Systems Analyze Water Quality in Real Time (WI)
27. Compact Radiometers Expand Climate Knowledge (MA)
28. Energy Servers Deliver Clean, Affordable Power (CA)
29. Solutions Remediate Contaminated Groundwater (LA)
30. Bacteria Provide Cleanup of Oil Spills, Wastewater (TX)
31. Reflective Coatings Protect People and Animals (CA)

Computer Technology

32. Innovative Techniques Simplify Vibration Analysis (AL)
33. Modeling Tools Predict Flow in Fluid Dynamics (PA)
34. Verification Tools Secure Online Shopping, Banking (CA)
35. Toolsets Maintain Health of Complex Systems (CT)
36. Framework Resources Multiply Computing Power (TX)
37. Tools Automate Spacecraft Testing, Operation (MD)
38. GPS Software Packages Deliver Positioning Solutions (CO)
39. Solid-State Recorders Enhance Scientific Data Collection (CO)
40. Computer Models Simulate Fine Particle Dispersion (NH)

Industrial Productivity

41. Composite Sandwich Technologies Lighten Components (OH)
42. Cameras Reveal Elements in the Short Wave Infrared (NJ)
43. Deformable Mirrors Correct Optical Distortions (MA)
44. Stitching Techniques Advance Optics Manufacturing (NY)
45. Compact, Robust Chips Integrate Optical Functions (MT)
46. Fuel Cell Stations Automate Processes, Catalyst Testing (WA)
47. Onboard Systems Record Unique Videos of Space Missions (CA)
48. Space Research Results Purify Semiconductor Materials (TX)
49. Toolkits Control Motion of Complex Robotics (MA)

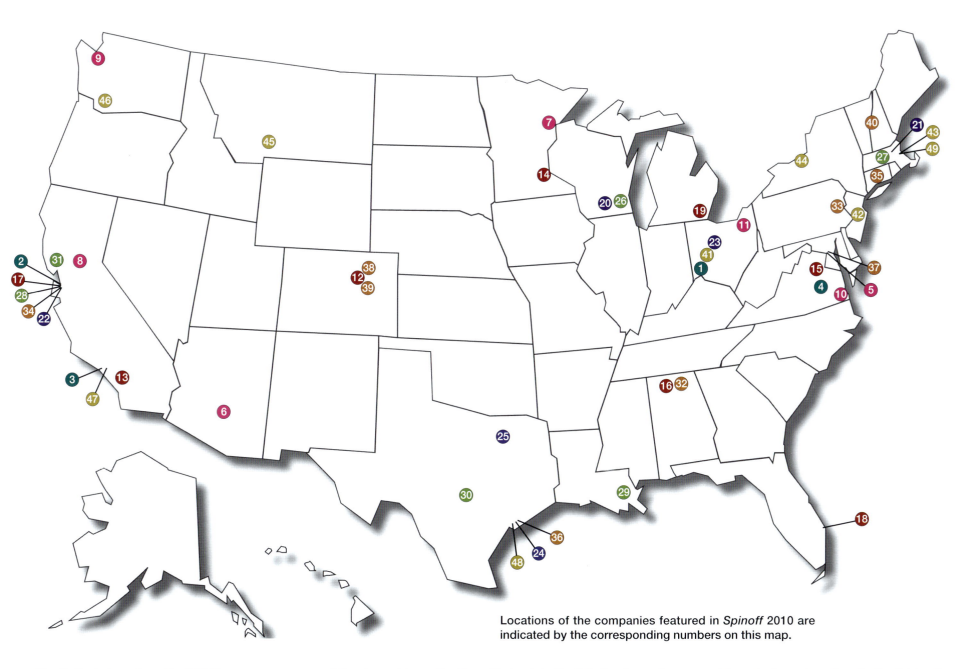

Locations of the companies featured in *Spinoff* 2010 are indicated by the corresponding numbers on this map.

The Nation's investment in NASA's aerospace research has brought practical benefits back to Earth in the form of commercial products and services in the fields of health and medicine; transportation; public safety; consumer goods; environmental resources; computer technology; and industrial productivity. *Spinoff*, NASA's premier annual publication, features these commercialized technologies. Since its inception in 1976, *Spinoff* has profiled NASA-derived products from companies across the Nation. An online archive of all stories from the first issue of *Spinoff* to the latest is available in an online database at www.sti.nasa.gov/spinoff/database.

NASA Technologies Benefiting Society

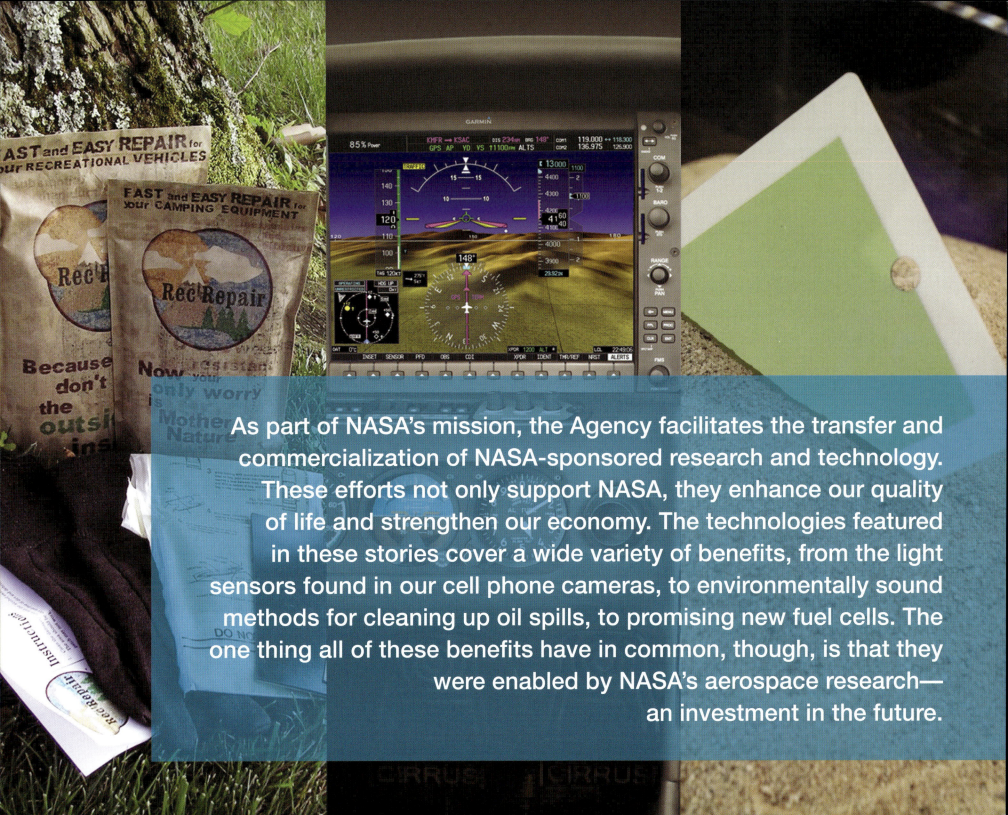

As part of NASA's mission, the Agency facilitates the transfer and commercialization of NASA-sponsored research and technology. These efforts not only support NASA, they enhance our quality of life and strengthen our economy. The technologies featured in these stories cover a wide variety of benefits, from the light sensors found in our cell phone cameras, to environmentally sound methods for cleaning up oil spills, to promising new fuel cells. The one thing all of these benefits have in common, though, is that they were enabled by NASA's aerospace research—
an investment in the future.

Health and Medicine

From cutting-edge rehabilitation devices to vision-saving lenses, NASA's research often results in technologies that keep us healthy—and even save our lives. The technologies featured in this section:

- Strengthen Hip Implants
- Monitor Cranial Pressure
- Protect, Enhance Vision
- Increase Imaging Capabilities

Burnishing Techniques Strengthen Hip Implants

Originating Technology/NASA Contribution

When compressed air mixes with jet fuel and is ignited in a turbine engine, the temperature can reach 3,000 °F. As a result of this fiery exhaust, the turbine spins and then forces the air through the back of the engine, and the jet moves forward. While extremely hot air assists in propelling a plane, it can also take a toll on the turbine blades and propeller hubs.

An engine component's lifespan is limited not only by heat, but also by general fatigue (weakening), corrosion, fretting fatigue (mechanical wear and oxidation that leads to cracking), and foreign object damage. Even a small amount of damage can cause a failure that can result in catastrophic consequences. Inspection and maintenance to avoid these failures in aerospace turbine engines is estimated to cost billions of dollars annually.

Techniques such as shot peening (impinging small steel spheres on a surface), laser shocking (using a laser to apply shock waves to a material), and deep-rolling (applying force by rolling a tool over the surface) are often used to apply compressive residual stress that actually boosts the strength of tough metal engine components.

In the 1990s, when NASA was looking for new and improved methods to increase the lifespan of engine components that undergo extreme temperatures and service, it found an alternative process called low plasticity burnishing (LPB), developed by Lambda Research Inc., of Cincinnati, Ohio.

Based on a series of studies on the thermal stability of a variety of surface treatments including shot peening and laser shocking, Lambda discovered that the more cold work (working of metal at room temperature) a material underwent, the less strength it retained when subjected to high temperatures. In developing LPB, Lambda used only a fraction of cold working, which increased the damage tolerance of materials and prevented cracking in components designed for high-temperature situations.

Partnership

To demonstrate LPB in the hot sections of turbine engine metal components, NASA's Glenn Research Center awarded Phase I and II **Small Business Innovation Research (SBIR)** contracts to Lambda in the late 1990s. Through these SBIRs, Lambda showed LPB to be an affordable means of producing a thermally stable deep layer of compressive residual stress in metallic components that remained stable at engine operating temperatures. LPB also increased the lifespan of components, doubled the endurance limit of components, halted existing cracks, and improved the fatigue performance of turbine alloys without altering the alloy or the design.

Performed by rolling a hydrostatic bearing tool over the surface of a specific part or piece, LPB allows an exact amount of force to create a desired layer of compression in one pass. By producing a repeatable and stable deep layer of compressive surface residual stress, metals become more resistant to corrosion, damage from foreign objects, and cracking.

"NASA gave us the initial opportunity to demonstrate LPB in an application that provided the new technology to the aircraft engine, ground-based turbine applications, and to aging aircraft. The initial NASA SBIR was also instrumental in supporting additional, more extensive funding that was available through the Department of Defense, primarily with the Naval Air Systems Command (NAVAIR) and some with the Air Force, which has led to the introduction of LPB into commercial aircraft, now with the support of the Federal Aviation Administration (FAA)," says Paul Prevéy, CEO of Lambda Technologies Group.

Developed by Lambda Research Inc. and refined with NASA's Glenn Research Center's funding, low plasticity burnishing (LPB) strengthens metal components, like the F100 aircraft blade shown here, by applying force to the surface of a material to produce a desired layer of compressive residual stress. The technique has proven effective for components that undergo extreme pressure, temperature, and stress.

The LPB process completely eliminated the occurrence of fretting, which leads to cracking, in the neck segment of a medical hip implant (above left). It also increased the fatigue strength of the implant by 40 percent and the lifespan by more than 100 times. Performed in a machine shop environment, in the field, and by using industrial robotic tools (above right), LPB can be applied to new or old metal components.

Prior to completing its work with NASA, Lambda patented the LPB process and created a spinoff company, Surface Enhancement Technologies LLC, to market LPB. In 2010, LPB earned recognition as one of the "R&D 100" (a list of the top 100 inventions of the year), granted by *R&D Magazine*.

Product Outcome

Capable of being applied to all types of carbon and alloy steel, stainless steel, cast iron, aluminum, titanium, and nickel-based super alloys, and many components with odd shapes or forms, LPB can be performed in a machine shop environment, in the field, and by using robotic tools. One important feature of the LPB application method is that it is highly controllable and can be validated to ensure that the process is applied to every part.

Over the last decade, LPB has been successful in completely eliminating fatigue failures in the first stage vanes on aircraft engines. Previously, these failures had resulted in the loss of several aircraft and crew. Lambda has processed over 45,500 of these vanes, and achieved process control exceeding Six Sigma. This processing is what led to FAA acceptance of LPB as a suitable process for both repair and alteration of commercial aircraft components. It is estimated that LPB will save the aircraft market over $10 million for the mitigation of stress corrosion cracking on just one landing gear application.

In 2009, Lambda and Delta Airlines announced an exclusive partnering agreement to use LPB for maintaining commercial aircraft components such as landing gear, propeller hubs, and turbine engine blades. According to the companies, the team effort provides opportunities to extend the life of aging aircraft, overcoming damage from foreign objects, stress corrosion cracking, and corrosion pitting damage mechanisms that are common throughout the commercial aircraft fleet.

In addition to having a significant impact on defense and aerospace components, LPB has also had a major impact on medical implant manufacturers. Lambda finds the savings for the medical and aircraft markets combined could reach numbers in excess of $100 billion.

A 2004 partnership between Lambda and Exactech, an orthopedic company that develops, manufactures, markets, distributes, and sells orthopedic implant devices and related surgical instrumentation, initiated the first commercial application of LPB to stop fretting in medical implants. Prior to working with Exactech, Lambda addressed a similar problem under SBIRs with NAVAIR, building on the work of NASA SBIRs, to apply LPB to stop fretting damage on the blade dovetail joint on jet engines.

In the case of the hip implant, fretting was occurring in a section of the hip stem due to severe cyclical stress, as every step taken by a patient represented a single loading and unloading cycle. Exactech explored a number of solutions to increase the performance of the implant, including laser peening and roller burnishing, but nothing compared to LPB, which improved the fatigue strength of the hip stem by more than 40 percent and increased the lifespan of the piece by more than 100 times. Data from the U.S. Food and Drug Administration confirmed that LPB completely eliminated the occurrence of fretting fatigue failures in modular hip implants. LPB has been integrated in the manufacturing process and applied to more than 3,400 hip implants.

Another application where LPB was chosen over laser peening was in eliminating residual tension in the final closure weld of the long-term nuclear waste storage containers for Yucca Mountain. The design review board for the containers unanimously selected LPB for the greater depth of compression, advantages in logistics, quality control, surface finish, and cost. The U.S. Department of Energy found LPB produced residual compression that exceeded the depth required for the surface to remain in compression for the 50,000-year design life of the containers. ❖

LPB™ is a trademark of Lambda Technologies.
Six Sigma®/SM is a registered trademark and service mark of Motorola Inc.

Signal Processing Methods Monitor Cranial Pressure

Originating Technology/NASA Contribution

When you think of a beating heart, you might assume it beats at regular intervals, but in actuality, velocity and pressure change with every beat, and the time interval between each beat is different. Now a NASA-developed technology is helping researchers understand blood flow and pressure in ways that may improve treatment for victims of brain injury and stroke.

Dr. Norden Huang, a scientist and mathematician at Goddard Space Flight Center, invented a set of algorithms for analyzing nonlinear and nonstationary signals that developed into a user-friendly signal processing technology for analyzing time-varying processes.

Efficient and adaptive, it can be used to analyze data sets for a wide variety of applications and even improves accuracy when used with linear and stationary signals. Dubbed the Hilbert-Huang Transform (HHT), it is a continuation and adaptation of early 20th century mathematician David Hilbert's work on signal analysis.

An advantage of HHT over other common signal analysis techniques, like Fourier transforms, is that it is more precise and accurate, capable of sharper filtering while preserving the integrity of the data. It is also flexible with the types of data it can analyze, since it does not require the data sets to be linear and stationary.

As an added bonus, this series of algorithms is relatively easy to implement and operate.

While NASA designed the technology for structural health monitoring and damage detection, like nondestructive testing of the space shuttle orbiters, the applications outside of the Agency are nearly limitless. For example, the HHT method can assist in the understanding of sound and vibrations for highway noise reduction, submarine design, and speech and sound recognition analysis.

It can also help in environmental analysis, like mapping land and water topography or water and wind dynamics. Industrial applications include machine monitoring.

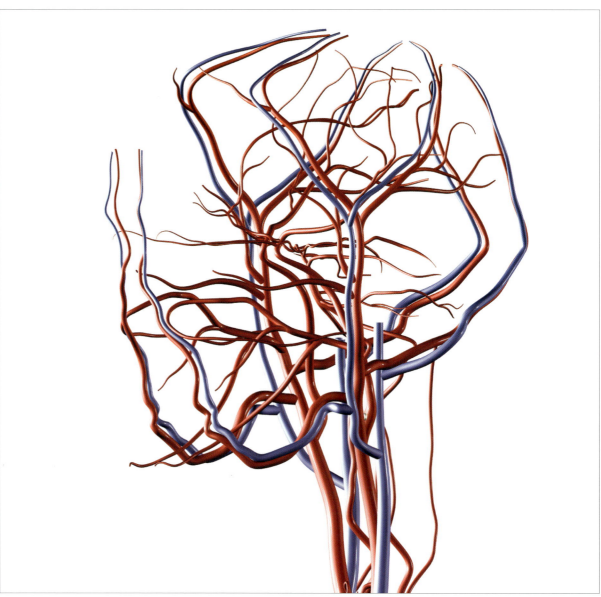

Medical professionals can use the data from Multimodal Pressure-Flow to create a reliable index of cerebral autoregulation and to help identify impairment of cerebral vasoreactivity, which is caused by medical syndromes that affect the brain, such as stroke, dementia, and traumatic brain injury. Shown here is the complex network of veins and arteries that carry blood to and from the brain.

The work on this project led to a Federal Laboratory Consortium "Technology Leadership Award" for the HHT method in 2000, as well as recognition as one of *R&D Magazine's* "R&D 100" (a list of the top 100 inventions of the year) in 2001, and the "NASA Government Invention of the Year" in 2003.

Partnership

When NASA was founded, the U.S. Congress required that the Agency work to make its technologies available to the public. Traditionally, NASA has made public announcements of patents available for license, either through news releases or publication in the monthly *Tech Briefs* magazine.

Recently, Goddard experimented—successfully—with a new approach to transferring NASA technology to the public by placing a handful of licenses to its patents up for auction. At an auction managed by Ocean Tomo Federal Services LLC, an intellectual property auction house, licenses of 10 U.S. patents and 1 domestic patent application were sold to DynaDx Corporation, of Mountain View, California, a medical technology incubator that develops and markets products to improve clinical diagnostics and prediction of medical problems. This lot of patents included the HHT.

Product Outcome

DynaDx is now using the licensed NASA technology for medical diagnosis and prediction of brain blood flow-related problems, such as stroke, dementia, and traumatic brain injury.

Through the course of normal, everyday activities, blood in the brain is shifted around to different sections, according to where it is needed for that activity. When a person suffers from one of these brain blood flow-related problems, the body often will not regulate blood flow to the brain, and this can manifest as cognitive impairment.

For example, a person who has suffered a traumatic brain injury may not be able to complete two tasks at the same time, even something as commonplace as tying shoes and holding a conversation simultaneously. The brain's ability to respond to these daily activities, providing a relatively stable blood flow during regular activities that may raise or lower blood pressure, is called dynamic cerebral autoregulation, and it is toward understanding this phenomenon that DynaDx is applying the NASA-derived HHT technology. DynaDx calls its HHT-based process Multimodal Pressure-Flow (MMPF), and it is proving to be an accurate and sensitive analysis of dynamic cerebral autoregulation.

One of DynaDx's research advisors, Dr. Vera Novak, a gerontologist with Beth Israel Deaconess Medical Center and an associate professor at Harvard University's School of Medicine, is using MMPF to study the effects of ageing on cerebral blood flow regulation, the body's ability to control the relationship between blood pressure and blood flow in the brain.

"After we had done several analyses, we found that the MMPF is more sensitive and specific than the current methods that assess autoregulation," Novak says. This work on geriatric patients can also be used to help researchers and doctors know more about other brain blood flow disorders, like those stemming from brain injuries and stroke.

According to Novak, traditional methods for analyzing blood pressure "presume that heart rate and blood pressure are at constant intervals. If you record a beat-to-beat signal of blood flow velocity, of blood pressure, there's a lot of information in these signals. They are not stable. They change with every beat, and also the time interval between heart beats is different, as is the blood pressure and blood flow velocities. Using MMPF really improves the sensitivity and specificity of the measurements."

MMPF is currently being used in research settings, but it could soon be moved into clinical applications where different methods could be tested to improve patient conditions, and MMPF could help determine which methods are effective. The technology also has potential application in acute care settings.

Clinicians could predict changes in intracranial pressure in patients, itself an important indicator of neurological status. And since the technology is non-invasive and simple to use, it can be employed in triage settings, alerting caregivers of changes in intracranial pressure in stroke or brain injury victims who would not otherwise qualify for more invasive methods. This can result in the patient being sent to the operating room earlier if necessary, a potentially lifesaving decision.

In addition to MMPF, DynaDx has developed the scientific computation software, DataDemon, which is the only commercial data analysis software with built-in HHT algorithm.

DataDemon software is very easy to use; no programming skill is required. Users can build the data analysis diagram with several simple mouse clicks, and the results are ready for viewing and final reporting.

Besides the HHT algorithm, DataDemon includes tools for data filtering, math calculation, statistics, matrix operation, data transformation, and other popular time-frequency analysis methods, such as short-term Fourier transform, (Enhanced) Morlet wavelet, and Hilbert transform. DataDemon software is available now in both academic and professional versions. ❖

> *The license for HHT was the first-ever auctioned Federal intellectual property.*

Ultraviolet-Blocking Lenses Protect, Enhance Vision

Originating Technology/NASA Contribution

In the 1980s, Jet Propulsion Laboratory (JPL) scientists James Stephens and Charles Miller were studying the harmful properties of light in space, as well as that of artificial radiation produced during laser and welding work. The intense light emitted during welding can harm unprotected eyes, leading to a condition called arc eye, in which ultraviolet light causes inflammation of the cornea and long-term retinal damage.

To combat this danger, the JPL scientists developed a welding curtain capable of absorbing, filtering, and scattering the dangerous light. The curtain employed a light-filtering/vision-enhancing system based on dyes and tiny particles of zinc oxide—unique methods they discovered by studying birds of prey. The birds require near-perfect vision for hunting and survival, often needing to spot prey from great distances. The birds' eyes produce tiny droplets of oil that filter out harmful radiation and permit only certain visible wavelengths of light through, protecting the eye while enhancing eyesight. The researchers replicated this oil droplet process in creating the protective welding curtain.

The welding curtain was commercialized, and then the scientists focused attention on another area where blocking ultraviolet light would be beneficial to the eyes: sunglasses. In 2010, the groundbreaking eyewear technology was inducted into the Space Foundation's Space Technology Hall of Fame, which honors a select few products each year that have stemmed from space research and improved our lives here on Earth.

Partnership

SunTiger Inc.—now Eagle Eyes Optics, of Calabasas, California—was formed to market a full line of sunglasses based on the licensed NASA technology that promises 100-percent elimination of harmful wavelengths and enhanced visual clarity. Today, Eagle Eyes sunglasses are worn by millions of people around the world who enjoy the protective and vision-enhancing benefits.

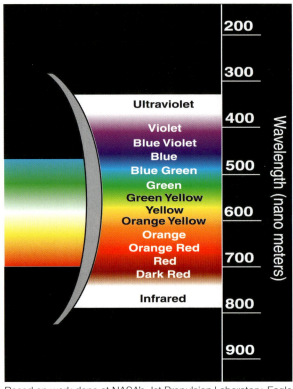

Based on work done at NASA's Jet Propulsion Laboratory, Eagle Eyes lenses filter out harmful radiation, reduce light scattering, and permit vision-enhancing wavelengths of light, protecting eyesight while also improving visibility.

Product Outcome

Maximum eye protection from the Sun's harmful ultraviolet rays is critical to our ability to see clearly. This is because when light enters the eye, a series of events happen which can help, hinder, or even destroy our eyesight. First, light passes through the cornea and ultimately reaches the retina which contains two types of cells—rods (which handle vision in low light) and cones (which handle color vision and detail). The retina contains 100 million rods and 7 million cones. The outer segment of a rod or a cone contains the photosensitive chemical, rhodopsin, also called "visual purple." Rhodopsin is the chemical that allows night vision, and is extremely sensitive to light. When exposed to a full spectrum of light, rhodopsin immediately bleaches out, and takes about 30 minutes to fully regenerate, with most of the adaption occurring in the dark within 5 to 10 minutes. Rhodopsin is less sensitive to the longer red wavelengths of light and therefore depleted more slowly (which is why many people use red light to help preserve night vision). When our eyes are exposed to the harmful, ultraviolet light rays of the Sun (UVA, UVB, and blue-light rays), damage to our eyes and their complex vision-enhancing processes can occur and not even be noticed until years later, long after exposure.

The most common form of eye damage related to ultraviolet exposure, cataracts, causes the lens of the eye to cloud, losing transparency and leading to reduced vision and, if left untreated, blindness. In the United States alone, it is estimated that cataracts diminish the eyesight of millions of people at an expense of billions of dollars. Other forms of eye damage directly attributable to ultraviolet exposure include pterygium, an abnormal mass of tissue arising from the conjunctiva of the inner corner of the eye; skin cancer around the eyes; and macular degeneration, which damages the center of the eye and prevents people from seeing fine details.

Alan Mittleman, president and CEO of Eagle Eyes explains, "When we're born, our eyes are clear like drops of water. Throughout life, we start to destroy those sensitive tissues, causing the yellowing of the eyes and the gradual worsening of eyesight. When the eye becomes more and more murky, cataracts form. Simple protection of the human eye, from childhood and throughout adulthood could protect the clarity of the eye and extend good vision for many years—even our entire lifetime.

The Eagle Eyes lens (right) makes scenes more vivid because harmless wavelength colors such as red, orange, yellow, and green are enhanced, and damaging rays in the blue, violet, and ultraviolet (UV) wavelengths are blocked.

"It has only been recently," he adds, "that people started to realize the importance of this." Sunglass manufacturers are recognizing the importance of eye care, and consumers are becoming more aware of eye health. One issue still plaguing the sunglass market, though, is that consumers assume that darker lenses are more protective, which is not always the case.

It may feel more comfortable to wear the dark lenses, but in addition to reducing the field of vision, it relaxes the eye, which allows more blue light to get directly to the retina. Blue light, in particular, has long-term implications, because it passes through the cornea and damages the inner retinal area.

The Eagle Eyes lens allows wearers to see more clearly because it protects from ultraviolet light, but more importantly, blocks this blue light, allowing the good visible light while blocking the harmful wavelengths.

Among their many donations throughout the years and goal of spreading good vision and eye protection to remote areas of the world, Eagle Eyes Optics had the opportunity recently to provide assistance to a group in sore need of eye protection: children in Galena, Alaska. The incidence of cataracts is 300 times greater in Alaska because of the Sun's reflection off of the snow. Eagle Eyes donated 150 pairs of its sunglasses to a high school in Galena, and they were delivered by members of the Space Foundation and presented by former NASA astronaut Livingston Holder. ❖

Hyperspectral Systems Increase Imaging Capabilities

Originating Technology/NASA Contribution

While the human eye can see a range of phenomena in the world, there is a larger range that it cannot see. Without the aid of technology, people are limited to seeing wavelengths of visible light, a tiny range within the electromagnetic spectrum. Hyperspectral imaging, however, allows people to get a glimpse at how objects look in the ultraviolet (UV) and infrared wavelengths—the ranges on either side of visible light on the spectrum.

Hyperspectral imaging is the process of scanning and displaying an image within a section of the electromagnetic spectrum. To create an image the eye can see, the energy levels of a target are color-coded and then mapped in layers. This set of images provides specific information about the way an object transmits, reflects, or absorbs energy in various wavelengths.

Using this procedure, the unique spectral characteristics of an object can be revealed by plotting its energy levels at specific wavelengths on a line graph. This creates a unique curve, or signature. This signature can reveal valuable information otherwise undetectable by the human eye, such as fingerprints or contamination of groundwater or food.

Originally, NASA used multispectral imaging for extensive mapping and remote sensing of the Earth's surface. In 1972, NASA launched the Earth Resources Technology Satellite, later called Landsat 1. It had the world's first Earth observation satellite sensor—a multispectral scanner—that provided information about the Earth's surface in the visible and near-infrared regions. Like hyperspectral imaging, multispectral imaging records measurements of reflected energy. However, multispectral imaging consists of just a few measurements, while hyperspectral imaging consists of hundreds to thousands of measurements for every pixel in the scene.

In 1983, NASA started developing hyperspectral systems at the Jet Propulsion Laboratory. The first system,

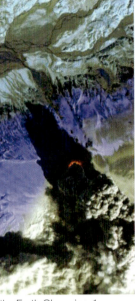

The Hyperion instrument onboard the Earth Observing-1 spacecraft obtained these images of Iceland's Eyjafjallajökull volcano. The left-hand image was created with visible wavelengths; the right-hand picture is an infrared image.

the Airborne Imaging Spectrometer, led to the development of the powerful Airborne Visible/Infrared Imaging Spectrometer (AVIRIS) that is still in use today. AVIRIS is connected to the outside of aircraft and is used to gather information to identify, measure, and monitor the environment and climate change. In 2001, NASA launched the first on-orbit hyperspectral imager, Hyperion, aboard the Earth Observing-1 spacecraft.

Partnership

Based on the hyperspectral imaging sensors used in Earth observation satellites, NASA engineers from Stennis Space Center and researchers from the Institute for Technology Development (ITD) collaborated on a new design that was smaller and incorporated a scanner that required no relative movement between the target and the sensor. ITD obtained a patent for the technology and then licensed it to a new company called Photon Industries Inc. In 2005, Lextel Intelligence Systems LLC, of Jackson, Mississippi, purchased the company and its NASA-derived technology (*Spinoff* 2007).

Without the technical expertise to market the product, the company's license for the scanner returned to ITD. In 2008, Themis Vision Systems LLC, of Richmond, Virginia, obtained an exclusive license for the technology. The CEO of Themis, Mark Allen Lanoue, was one of the original researchers on the staff that developed the device at ITD and saw the potential for the technology. In 2005, Lanoue, several colleagues, and the technology were inducted into the Space Technology Hall of Fame, created by the Space Foundation, in cooperation with NASA, to increase public awareness of the benefits that result from space exploration programs and to encourage further innovation.

Product Outcome

Themis delivers turnkey solutions in hyperspectral hardware, software, and algorithm development. Worldwide, Themis has built about 40 custom systems, including 3 at the Federal Bureau of Investigation's hyperspectral imaging laboratory in Quantico, Virginia. With distributors and customers in more than 10 countries, Themis recently installed the first UV hyperspectral system in China to help with studies in forensic science, including fingerprint analysis.

The latest product lines from Themis include the Transluminous Series, the Optoluminous Series, and HyperVisual Software. What is most unique about these hyperspectral systems is their size. Themis has developed compact, 4-pound systems—as opposed to the larger, 7- and 10-pound ITD versions—that can fit on multiple platforms including microscopes, tripods, or production lines.

The Transluminous Series spans the spectrum from UV to infrared and uses a prism-grating-prism component to split incoming light into separate wavelengths. The

Optoluminous Series, the newest line of reflective hyperspectral systems, uses a convex grating reflective spectrograph to gather wavelengths. Both lines feature the NASA-derived scanning technique that requires no relative movement between the target and the sensor.

The HyperVisual Software is a graphical user interface-based software package for end-user communication and control. Designed for the Windows operating system, it converts scanned images into a single image format containing spatial and spectral information. HyperVisual acquires and pre-processes the data so that it is ready for analysis. A number of pre-processing routines can be run on the images and then ported into off-the-shelf image processing packages.

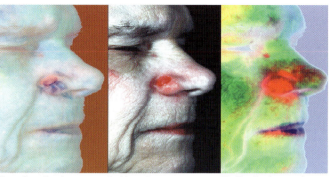

Based on the hyperspectral imaging sensors used in Earth observation satellites, NASA engineers and Institute for Technology Development researchers collaborated on a portable design for imaging the health of farmers' crops. Exclusively licensed by Themis Vision Systems LLC, the company has developed compact, 4-pound systems that can fit on multiple platforms (right). One medical application for the technology is imaging for skin cancer, cervical cancer, and breast cancer surgery. The images above show cancer on a patient's nose.

Early on, the primary application for hyperspectral imaging was for remote sensing for agricultural and land-use planning applications. Now the number of applications continues to grow in a variety of areas: medical and life sciences, defense and security, forensics, and microscopy.

Used in medical applications as a diagnostic tool, hyperspectral imaging looks at wounds and burns to monitor healing, scans skin to detect and monitor diseases, and looks inside eyes for diabetic retinopathy and clinically significant macular edema. In forensics, hyperspectral imaging examines ink colors to reveal counterfeit passports, currency, and checks. Microscopy applications include cell, spore, and DNA analysis.

The U.S. Department of Agriculture used Themis systems for imaging poultry, beef, and other food products. For poultry, the system captured an image of a bird and then processed the image to determine if the bird had a defect such as a skin bruise, tear, or fecal contamination. The imaging systems also produced spectral signatures of dirt, fungi, fecal matter, and pathogens such as *Salmonella* and *E. coli*.

A company called X-Rite uses a Themis hyperspectral system to assist in quality control and color measurement in paint mixing. Estée Lauder has utilized a Themis system to improve cosmetics and makeup coverage. One organization is even using the product to develop camouflage, while another is using it to detect camouflage. Other military and defense applications include detecting landmines, tripwires, and for search and rescue operations.

One of the more unusual applications for hyperspectral imaging is to get a closer look at paintings. "In art forensics, you look at paintings to see if they have hidden signatures. A lot of times, artists would paint over the signatures. Hyperspectral imaging helps to get a look underneath the layers of paint," says Lanoue.

In fact, Lanoue recently assisted a man with a painting that had been bought at an auction for $3,000. One day, the man noticed a faint signature under the obvious signature. Using a Themis hyperspectral imaging system, Lanoue was able to see the signature underneath. Lanoue and the owner are now working with art experts to confirm what the hyperspectral imaging seems to reveal: that the painting is actually the work of famous Spanish artist Diego Velázquez. This fact could help the painter's owner turn a significant profit.

Besides using hyperspectral imaging for such exciting new applications, Lanoue has filed for new patents for next-generation scanning systems and a biofuel sensory system. He also plans to continue selling turnkey hyperspectral imaging systems, including microimaging systems. Furthermore, Lanoue is working to release a future line of intelligent imaging systems and real-time hyperspectral applications based on the NASA-derived scanning technique. ❖

Transluminous™ and Optoluminous™ are trademarks of Themis Vision Systems LLC.

HyperVisual Image Analyzer® is a registered trademark of the Institute for Technology Development.

Windows® is a registered trademark of Microsoft Corporation.

Transportation

From pioneering the development of personal air vehicles to making our existing modes of transportation greener and more efficient, NASA technologies are often part of the ways we get around. The technologies featured in this section:

- Model the Future of Air Traffic Management
- Stabilize Helicopters, Reduce Noise
- Point to the Future of Transportation
- Lead to Potential New Vehicles
- Save Billions of Dollars in Fuel Costs
- Collect Critical Aerodynamics Data
- Extend Life of Engines and Infrastructure

Programs Model the Future of Air Traffic Management

Originating Technology/NASA Contribution

Try describing the U.S. National Airspace System (NAS), and you will inevitably end up rattling off a series of large numbers. There are more than 87,000 flights—commercial, general aviation, military, chartered, cargo—every day; about 5,000 flights in the air at any given moment; and more than 14,000 air traffic controllers working to manage the safety of all of these flights, including an average of 64 million takeoffs and landings a year (more than 7,000 every hour). In addition, there are more than 19,000 airports; 600 air traffic control facilities; more than 70,000 radar systems, communications relays, and other equipment; and thousands of technicians and safety inspectors. The NAS includes all of these components and others, like the 660 million passengers and 37 billion cargo-revenue tons of freight that crisscross the Nation every year. Even the weather is considered part of the NAS.

If all of those numbers are not mind-boggling enough, consider that airspace demand is expected to as much as triple by 2025. The increase in air traffic is outpacing construction of airports and runways and, of course, the actual amount of airspace available remains unchanged. Crowded skies promise safety issues, longer delays and more congestion at airports, and less response time for air traffic controllers managing the flow. In addition, the NAS has become an increasingly interconnected and interdependent system: a snowstorm grounding flights in the Northeast, for example, leads to disruptions, delays, and economic impacts nationwide.

In keeping with its mission to "enable a safer, more secure, efficient, and environmentally friendly air transportation system," NASA's Aeronautics Research Mission Directorate is working closely with the Federal Aviation Administration (FAA) to develop technologies and procedures for creating a Next Generation Air Transportation System, transforming the entire NAS to address these problems.

Partnership

Evaluating potential air traffic management (ATM) solutions in real life, however, is difficult and costly; computer models and simulations can provide far more practical and comprehensive alternatives. But modeling and simulating a system with the scale and complexity of the NAS in useful detail is a daunting task, and doing so in a way that allows for the flexible testing of ATM scenarios adds further challenges.

To help develop a program capable of simulating a realistic NAS down to individual flights, NASA partnered with Rockville, Maryland-based Intelligent Automation Inc. (IAI). Through **Small Business Innovation Research (SBIR)** contracts with Ames Research Center, IAI continued work on specialized software the company had begun developing with funding from the U.S. Department of Defense (DoD). With the NASA support, IAI produced the software, called Cybele, and released it as an open-source program. The unique capabilities of the Cybele program led to its inclusion as an enabling technology for NASA's Airspace Concept Evaluation System (ACES), which is providing researchers with glimpses of the future of the national airspace.

Product Outcome

Cybele is an infrastructure for autonomous, interactive components called agents, which researchers program to model the behaviors of various entities—such as airplanes or even individual people in a crowd. The agents can operate while distributed across multiple computers and execute their activities in individual threads, interacting

With more than 87,000 flights every day and airspace demand that is expected to triple by 2025, solutions are needed to avoid longer delays and more congestion at airports.

with each other via messaging. These qualities make Cybele an excellent framework for modeling complex systems with multiple, individual constituents—like the NAS.

"In the ATM domain, an agent can correspond to an aircraft, while another agent corresponds to an air traffic controller, and the communications between the controller and the aircraft are modeled via messages," explains Vikram Manikonda, president of IAI.

Cybele currently forms the agent infrastructure for ACES, creating agents for every part of the NAS, down to individual airplanes, airports, control centers, and even models of en route winds and other meteorological elements. The agents generate data streams that mimic the interactions of their real-world counterparts, and ACES collects the data for analysis, allowing researchers to determine how the NAS will react to increasing airspace demand and the "ripple effect" of disruptive elements like bad weather. Cybele provides ACES with the flexibility to model anything from a single airport or control center up to the entire NAS in fast time, and its distributed operation maximizes available computational resources. Though still being expanded and improved, ACES is currently used in a number of studies, including NASA's Advanced Airspace Concept, an effort to increase airspace capacity.

IAI serves as a subcontractor for technology innovator Raytheon on the ACES contract and has licensed Cybele to a number of aerospace companies and NASA contractors, as well as NASA's Ames and Langley Research Centers. The company now offers a range of Cybele products, including the company's core agent programming product CybelePro; the complete agent-based development platform CybelePro Enterprise; and the more limited, open-source OpenCybele and free CybeleLite versions.

Cybele's capabilities are not limited to ATM research. Other companies have licensed the software for applications in distributed robotics, and the U.S. Air Force has used Cybele to model the behavior of crowds.

"We modeled crowds as agents, simulating aggressive crowds, passive crowds, what influence a leader would have on crowd behavior, and how crowds would disperse in certain situations," Manikonda says. IAI provided the same agent-based modeling for the U.S. Army to study leadership development and troop training.

Over the past two decades, IAI has benefited from several SBIR contracts with NASA. In fact, Manikonda says IAI's SBIR-derived Cybele technology was instrumental in the formation of the company's ATM group that currently employs 20 people.

Ames Research Center's Future Flight Central simulator, the world's first full-scale virtual airport control tower, is just one of many ways NASA tests solutions to solve potential air and ground traffic problems at commercial airports.

"We view NASA partnership as an integral part of our growth," says Manikonda. "We've used the SBIR program as seed money to grow innovative ideas."

As IAI continues to work on SBIR-funded endeavors, including ATM projects focusing on concepts such as dynamic airspace corridors and airborne merging and spacing, the outcomes of these efforts, the ACES project, and other NASA partnerships may well determine the health of the Nation's airspace in the coming decades. ❖

Cybele™ is a trademark of Intelligent Automation Inc.

Tail Rotor Airfoils Stabilize Helicopters, Reduce Noise

Originating Technology/NASA Contribution

Given the eye-catching nature of space shuttle launches, deep-space imagery, and Mars exploration, it can be easy to forget NASA's aeronautics efforts, which have a daily impact on life within the bounds of Earth's atmosphere. Virtually every flying vehicle in operation today has benefited in some way from NASA advancements, and the helicopter is no exception. In fact, NASA's involvement in rotorcraft research can be traced back to its predecessor, the National Advisory Committee for Aeronautics (NACA). NACA was founded in 1915, less than a decade after the first successful piloted rotorcraft flight in 1907, and made a number of contributions to rotorcraft development—including a series of airfoils that are still employed in some modern vehicles.

NASA was formed in 1958, and within a little more than a decade the Agency had begun a collaborative rotorcraft research program with the U.S. Army, establishing laboratories at Ames Research Center, Glenn Research Center (then known as Lewis Research Center), and Langley Research Center. These labs focused on enhancing the performance and safety of helicopters for both military and civilian use. This research improved helicopter airfoil designs, flight control systems, aerodynamics, rotor blade and aircraft body composition, and cockpit configuration.

Partnership

Among the many outcomes of the NASA-Army research partnership are advanced airfoils designed and wind-tunnel tested at Langley. Two of these airfoil designs—the RC(4)-10 and the RC(3)-10—were licensed and commercialized by Carson Helicopters (*Spinoff* 2007) as a superior replacement main rotor for the Sikorsky S-61 helicopter, allowing the helicopter to fly faster and carry heavier loads while offering a service life twice that of the original rotor.

Carson's success with its NASA-derived airfoil caught the eye of Dean Rosenlof, general manager and aerospace engineer for Van Horn Aviation LLC (VHA), based in Tempe, Arizona. The company—founded by former Ames engineer Jim Van Horn, who worked on NASA rotorcraft research like the Rotor Systems Research Aircraft in the early 1980s—was looking for airfoil designs to expand its tail rotor blade product offerings, which include an aftermarket carbon composite tail rotor for the UH-1H ("Huey") military helicopter. Rosenlof brought the Langley RC series of airfoils—the low-speed RC(4)-10 and the high-speed RC(5)-10—to Van Horn's attention, and they determined these were precisely the designs the company was looking for.

The complex aerodynamics of a helicopter present a challenge to airfoil designers, who must consider a range of aerodynamic forces and how they influence the rotorcraft's flight capabilities. Among the chief concerns is pitching moment, the twisting force exerted by the airfoil that pushes the noses of the rotor blades up or down. Because this force can interfere with pilot control and rotor blade stability, designers aim to create airfoils with minimal to zero pitching moment.

"The RC airfoils were exactly what we needed," says Van Horn. "They are very attractive in that they are thin, light, laminar-flow airfoils with essentially zero pitching moment."

VHA contacted Langley and discovered that the airfoils' patent had expired, meaning the original NASA designs had entered the public domain.

"Langley encouraged us to take the designs, go forward, and be fruitful," says Van Horn.

Product Outcome

A helicopter tail rotor serves two essential functions. It provides a counteracting force to the helicopter's main rotor; without the sideways thrust produced by the tail rotor, the torque generated by the main rotor would spin the helicopter's body in the opposite direction. The tail rotor also allows the pilot to steer the helicopter around its vertical axis by adjusting the pitch of the rotor blades. Using the design for the NASA RC(4)-10 airfoil, VHA crafted an updated aftermarket tail rotor for the popular Bell 206 series of helicopters.

"It's an excellent airfoil, very stable, with very high stall margins," says Van Horn. The company built upon the RC(4)-10 airfoil, employing corrosion-resistant composite material with a titanium root fitting, a swept tip, a nickel abrasion strip that reduces wear on the blades' leading edges, and a new pitch bearing design. The result is a highly durable tail rotor blade—the Federal Aviation Administration (FAA) granted the VHA 206 tail rotor a 5,000-hour lifetime, twice that of the original equipment

The Rotor Systems Research Aircraft (RSRA) is seen here in flight over Ames Research Center. Former Ames engineer Jim Van Horn, founder of Van Horn Aviation (VHA), worked on the RSRA in the early 1980s.

VHA's NASA-derived 206 tail rotor is made of composite material with additional features like a titanium root fitting, swept tip, nickel abrasion strip, and new pitch bearing design.

Employing the NASA-developed RC(4)-10 airfoil design, the VHA 206 rotor blade provides a high-performance aftermarket option for the popular Bell 206 series helicopter.

manufacturer blade—with a number of enhanced features. The airfoil possesses zero pitching moment within typical operating speeds, Van Horn explains, and while the NASA airfoil's design already limits the turbulence that causes noise and drag, the inclusion of the swept tip further reduces these undesirable qualities. FAA-mandated testing demonstrated a 40-percent reduction in the overall sound exposure level (the amount of noise produced) for helicopters employing the VHA 206 tail rotor—a welcome improvement for pilots, passengers, and people on the ground. In addition, the airfoil's high stall margins enhance helicopter performance at high altitude; VHA flew helicopters with the new tail rotor at the Leadville Airport in Colorado, the highest elevation airport in North America, and determined the NASA-derived blade delivered superior high-altitude performance compared to the existing model. These improvements stand to benefit helicopter performance for a wide range of missions, including law enforcement and homeland security, military training, aerial patrol of wildfires and pipelines, mosquito control, and emergency medical services.

The tail rotor received FAA certification in 2009, and VHA delivered its first shipment of the NASA-derived blades to customers that same year. Now the company plans to use the NASA airfoils as its go-to design for all future projects aimed at advancing rotorcraft performance, Van Horn says. He adds that VHA has helped ensure a solid base for its future by taking advantage of NASA research.

"Given the market size and that we could capture a reasonable market share, this puts our company on very firm footing for the next 10 to 20 years and will provide a steady income to allow us to grow at a reasonable rate and develop new products," he says.

"I've been on both sides of the government-research-to-commercial-product equation, and it's a great system. It gives us an advantage that other companies don't have, because we were able to avail ourselves of this NASA technology." ❖

Personal Aircraft Point to the Future of Transportation

Originating Technology/NASA Contribution

In the late 1970s, general aviation (GA) in the United States was experiencing its heyday. In 1978, as many as 18,000 GA aircraft were produced. But only 15 years later, the industry was on the verge of collapse, with fewer than 1,000 aircraft produced in 1993.

One of the reasons for this decline was the lack of technological development that exposed the industry to safety and efficiency concerns. NASA, however, saw great potential within the GA industry to revolutionize the U.S. transportation system. With congestion growing both on the roads and in the skies, the Agency envisioned a Small Aircraft Aviation System, or SATS, in which improved GA aircraft would serve as an efficient travel option for round-trip distances too long to comfortably drive but too short to be practical for regular commercial airline service. Making use of the Nation's 19,000 airports (of which 14,000 are privately operated), SATS would provide an alternative to crowded highways and the overburdened hub-and-spoke airline system.

In order to facilitate the creation of SATS, GA aircraft needed to become cheaper to produce, quieter and more fuel efficient, and easier and safer to fly. In 1994, NASA and the Federal Aviation Administration (FAA) joined with private industry, academia, and nonprofits to form the Advanced General Aviation Transport Experiments (AGATE) consortium. Consisting of about 70 members

Originally built to allow Apollo astronauts to practice landing on the Moon, this 240-foot-high gantry (left) is now Langley Research Center's Landing and Impact Research Facility. Crashworthiness testing at the facility (right) allowed Cirrus Design Corporation to improve survivability during stall/spin impacts.

and led by NASA's Langley Research Center, AGATE sought to revitalize the GA industry and help drive the technological innovation needed to make SATS viable. Among the technologies AGATE focused on were safety and crashworthiness improvements, guidance systems, aerodynamically efficient airfoils, and manufacturing processes.

While AGATE ended in 2001, NASA continues to inspire the GA industry through efforts like the upcoming 2011 Green Flight Challenge, which seeks to demonstrate personal aircraft featuring maximized fuel efficiency, improved safety, and reduced noise. In the meantime, innovations with origins in the AGATE program continue to shape general aviation today.

Partnership

Among the primary tools AGATE employed to stimulate innovation and technology transfer were the **Small Business Innovation Research (SBIR)** and **Small Business Technology Transfer (STTR)** programs. Companies that received Phase II SBIR or STTR contracts were invited to join the AGATE consortium, further encouraging collaboration among government, academic, and industry partners.

One such company was Cirrus Design Corporation, based in Duluth, Minnesota. Founded in 1984, Cirrus' first product was an experimental aircraft, the VK-30. The company was keen on improving personal aircraft performance by making use of natural laminar flow; vehicles that take advantage of this property experience significantly less drag and thus fly faster and with better fuel efficiency. The VK-30 featured a natural laminar flow airfoil (the NLF-414F) designed by Langley engineer Jeff Viken. The problem with manufacturing the airfoil, however, was that production methods that used aluminum to craft the wing ultimately destroyed the laminar flow properties of the airfoil.

Through SBIR contracts with Langley, Cirrus worked on developing low-cost manufacturing methods using composite materials, which would provide a strong, lightweight aluminum alternative that preserved natural laminar flow. At the time, composites were used either for boats or for high-end military aircraft, says Cirrus chief engineer, Paul Johnston.

"We needed the composites to be aerospace quality but more in line cost-wise with what it would take to make a boat," he says. Cirrus' SBIR work resulted in significant composite manufacturing expertise and a pre-impregnated composite that could be readily mass-produced.

Additional SBIR work with Glenn Research Center explored electroexpulsive deicing systems to help ensure safe operation in dangerous icing conditions. While this technology ultimately proved commercially impractical for Cirrus, the company developed a method under the SBIR for mating the system to aircraft wings without disrupting their natural laminar flow. Cirrus later applied the same method to install glycol "weeping wing" systems, providing chemical icing protection without sacrificing performance.

Product Outcome

Cirrus employs both of these SBIR-derived benefits in the production of its industry-leading aircraft today. The Cirrus SR20 and the faster, more powerful SR22 and SR22 TURBO personal aircraft are currently among the most popular GA aircraft in the world; the SR22 has been the top-selling FAA-certified single-engine airplane every year since 2002.

Among the features that have earned Cirrus planes such popularity are a host of innovations with NASA

Cirrus' synthetic vision systems trace their roots to innovations in glass cockpit technology developed under the NASA-led Advanced General Aviation Transport Experiments consortium.

Featuring a host of NASA-derived design and technology features—including an additional NASA spinoff, the standard Cirrus Airframe Parachute System manufactured by BRS Aerospace Inc.—the Cirrus SR22 is the top-selling FAA-certified single-engine airplane in the world.

The Cirrus SR20 benefited from low-cost composite materials manufacturing that Cirrus developed through SBIR contracts with Langley.

connections. Perhaps the most important are the comprehensive safety features. GA aircraft typically fly at too low of an altitude to recover from a spin in time to avoid impact. NASA researchers in the 1970s and '80s focused on methods to help prevent aircraft from getting into a spin in the first place. Cirrus now employs a NASA-designed "drooped" leading edge on its airfoils that lowers stall speed and greatly increases spin resistance. In addition, through crashworthiness testing in the late 1990s—using Langley's Landing and Impact Research Facility, a 240-foot-high gantry originally built to train Apollo 11 astronauts for their historic Moon landing—Cirrus incorporated design features to improve survivability during stall/spin impacts.

"We also tested airbags," says Johnston. "There were no such thing as airbags in airplanes at the time, and now that's an option on all of our planes." Cirrus offers AmSafe Aviation Inflatable Restraints—seatbelt airbags—and was the first aircraft manufacturer to install the devices.

Another major safety feature is the Cirrus Airframe Parachute System, a whole-aircraft parachute capable of rescuing not only the pilot and passengers, but the entire plane. Invented by BRS Aerospace Inc. with NASA SBIR support (*Spinoff* 2002, 2010), the technology is standard on all Cirrus aircraft and has saved 35 Cirrus pilot and passenger lives to date.

Also key to enhancing safety are innovations that make Cirrus planes easier to fly.

"The AGATE program looked at how to take the essential information pilots need and display it to them in a manner that is intuitive and easy to use," says Johnston. One of AGATE's major contributions was the advancement of glass cockpit technology. (This does not refer to an airplane's windows. Rather, a "glass cockpit" features electronic instrument displays.) Among these technologies were synthetic vision systems that create three-dimensional renderings of the environment outside the aircraft, helping a pilot navigate, read the terrain, identify obstacles, and negotiate airborne traffic. One synthetic vision feature was the "Highway in the Sky," or HITS, a technology developed by avionics company Avidyne under a NASA contract. HITS simplified navigation by displaying boxes on the aircraft's screen that the pilot can (virtually) fly through—much like a video game. All of Cirrus' aircraft incorporate this technology, either through an Avidyne system or the new Cirrus Perspective synthetic vision system.

"This technology presents the pilot with the information necessary to fly without requiring the massive amounts of training and proficiency needed with the previous instrumentation," says Johnston. "It becomes easy to take on flight tasks that were once only for the highest experts."

The company's latest venture represents the newest wave of GA aircraft: the very light jet, or VLJ. Cirrus is developing its Vision personal jet, which the company promises will be lighter, quieter, and more efficient than other personal aircraft—another step toward making SATS a viable reality. The Vision is powered by the FJ33 turbofan engine, developed by engine manufacturer Williams International as part of the General Aviation Propulsion project, representing yet another legacy of the NASA-led AGATE program.

"NASA plays a role in looking at the transportation infrastructure as a whole and figuring out how to make it as efficient as possible to serve the most people with the least amount of resources," says Johnston.

"I don't think much of what you see in general aviation today would be around if NASA had not laid the foundation." ❖

Powered by a turbofan engine designed as part of the NASA General Aviation Propulsion project, the Cirrus Vision jet promises to provide the kind of quiet and efficient flight option that could make such vehicles viable, commonplace options for personal travel in the future.

AmSafe Aviation Inflatable Restraint® is a registered trademark of AmSafe Inc.

Cirrus Airframe Parachute System™ and Cirrus Perspective™ are trademarks of Cirrus Design Corporation.

Ducted Fan Designs Lead to Potential New Vehicles

Originating Technology/NASA Contribution

From the myth of Icarus, who flew too close to the Sun on wings made of wax, to the designs Leonardo da Vinci drew of flying machines that mirrored the wing patterns of birds, people have always dreamed of personal flight. In 1903, on a cold December morning in North Carolina, the Wright brothers made the dream a reality with the first manned flight. It lasted only 12 seconds, but initiated a rapid evolution in aircraft design, and within a few years there was an aircraft industry.

In the early days of manned flight, though, one idea persisted: the personal air vehicle. At the time, the concept of personal flight was still a thing of the imagination—epitomized in 1928 by the fictional character Buck Rogers, complete with his rocket belt.

In the 1950s, Bell Aerospace took the dream one step closer to reality with its unveiling of the jet belt, a small, low-thrust rocket that strapped to the operator's back. The short flight of 20 or 30 seconds, however, was not enough to make it viable for anything practical.

In 1955, with funding from the U.S. Navy, Hiller Aviation created the Hiller Flying Platform, a rotorcraft that was essentially a disk with a helicopter underneath. The operator stood on the platform and steered by shifting weight. Although there was interest and the prototype showed promise, the craft never went into production, as the standard helicopter proved more practical. For the next few decades, most of the interest in flight focused on the jet engine, and personal aircraft design was again relegated to the stuff of fiction.

For the centuries that people had dreamed of personal flight, there were countless great ideas, thousands of drawings, and hundreds of planned attempts. The only problem was that none of them stayed in the air long enough, so the dream lay dormant. But in the 1990s, new, lighter, stronger materials and advanced computer design systems awakened that dream.

The OVIWUN, a miniaturized version of Trek Aerospace Inc.'s larger, manned vehicles, allows researchers to test designs and experiment in a smaller, safer, more contained setting.

Partnership

In 1994, two aerospace engineers, Rob Bulaga and Mike Moshier, drew sketches for an aircraft they believed could prove viable, and by 1996, had formed a company, Trek Aerospace Inc. The company, based in Folsom, California, took full advantage of its proximity to NASA's Silicon Valley-based Ames Research Center for a great deal of testing, results of which have provided greater lift, lowered weight, more power, and improved maneuverability.

In 2000, using a wind tunnel at Ames, the engineers improved their designs. They tested their duct and fan system at the NASA site and were able to watch the flow of air over the ducts at various angles, finding that there was a very small stall area, and that for the most part, the flow did not separate. This clean airflow showed them that the craft was accomplishing 40 percent of its lift out of the duct system, which meant that the engineers could accomplish lift with a significantly smaller, lighter engine.

The experience gave the engineers a better understanding of how their craft worked and led to several design changes, including the use of a fly-by-wire system. The original prototype had handgrips and relied on the operator to shift his weight in order to operate the vehicle, but the wind tunnel testing suggested that this would not give the pilot adequate control of the vehicle. The fly-by-wire solution replaces the handgrips with two joysticks, one for controlling altitude and the other for turns. Information from the joysticks is fed into an onboard computer.

Product Outcome

All of Trek Aerospace's aircraft employ ducted, counter-rotating fans attached to a central gearbox and drive train, connected to a power source. The ducts allow the craft to fly into tight spaces without fear of damaging the rotors or anything else with which the rotors would otherwise come in contact. While seemingly simple, the company suggests that its success with the vehicles is the right combination of devices and how to make them interact effectively. The technology has been applied to three models: the Dragonfly UMR-1, the Springtail EFV, and the OVIWUN.

The Dragonfly UMR-1 (unmanned/manned/or remotely operable), the only horizontally configured craft the company has designed, is still in development at its onsite hangar. The company anticipates that civilian uses for the craft will include everything from crop dusting to commuting, and military uses will abound, whether as an unmanned reconnaissance vehicle or as its 450-pound payload capacity is leveraged to transport injured soldiers. It is available for purchase, but only as what the Federal Aviation Administration calls an experimental aircraft.

Who hasn't been stuck in traffic wishing that there was a way to rise above the throngs of other commuters?

The Springtail EFV (exoskeleton flying vehicle), currently in development, but with several working prototypes finished, uses a series of ducted propellers fueled by a 118-horsepower rotary engine. It fits pilots from 5 feet 4 inches to 6 feet 6 inches in height and weighing from 115 to 275 pounds. It has a top speed of 113 miles per hour and can soar up to 11,400 feet, though the designers intend for the vehicles to operate around 400 feet off the ground and to cruise at a comfortable 90 miles per hour.

The military is quite interested in this vehicle, and Trek Aerospace has received significant funding from the Defense Advanced Research Projects Agency (DARPA), the research and development arm of the U.S. Department of Defense. The vehicle has the potential for use by soldiers, as well as for search and rescue missions, reconnaissance, and surveillance. In addition, it has uses in the homeland security realm for firefighting, police work, and other emergency response situations.

Of course, there is also civilian interest. Who hasn't been stuck in traffic wishing that there was a way to rise above the throngs of other commuters? A personal aircraft would be the perfect solution. While the Springtail EFV is available for purchase as an experimental aircraft, Trek Aerospace is continuing to test it. The company, however, has had to develop an efficient method to continue the testing as well as procure additional funding.

Its solution was to make a miniaturized test version. Dubbed the OVIWUN, the small-scale version is for sale through the company's Web site and comes complete with a radio transmitter and receiver, battery charger, open source software, and a basic instruction manual. It weighs fewer than 6 pounds and can lift a little over that. This release has created a buzz among aerospace engineers and university computer science departments, the primary audiences.

The OVIWUN boasts two 450-watt electric motors that can deliver a maximum speed of 44 miles per hour and can climb 2,280 feet per minute. The radio-controlled craft's ducts allow for safer operation of the vehicle, as the rotors are protected and items that may be in their path are protected from them. The ducted covers allow the craft to bump against objects without damage, which brings to light the most significant advantages of this diminutive aircraft: It is safer than a manned vehicle, and its size makes it relatively difficult for it to damage itself during test flights the way a larger mass, faster craft could.

While this craft is not large enough to carry human passengers, it is definitely a sign of things to come. Someday soon, the dream of accessible individual civilian aircraft will likely be a reality. ❖

Springtail™ and OVIWUN™ are trademarks, and Dragonfly® is a registered trademark of Trek Aerospace Inc.

Trek Aerospace's Springtail EFV is a single operator powered-lift vehicle that supports the operator/pilot in a standing position.

Winglets Save Billions of Dollars in Fuel Costs

Originating Technology/NASA Contribution

Anyone who has made a paper airplane knows that folding the wingtips upward makes your plane look better and fly farther, though the reasons for the latter might be a mystery. The next time you snag a window seat on an airline flight, check out the plane's wing. There is a good chance the tip of the wing will be angled upward, almost perpendicular. Or it might bend smoothly up like the tip of an eagle's wing in flight. Though obviously more complex, these wing modifications have the same aerodynamic function as the folded wingtips of a paper airplane. More than an aesthetically pleasing design feature, they are among aviation's most visible fuel-saving, performance-enhancing technologies.

Aerodynamics centers on two major forces: lift and drag. Lift is the force that enables a plane to fly. It is generated by unequal pressure on a wing as air flows around it—positive pressure underneath the wing and negative pressure above. Drag is the resistance encountered while moving through the airflow. A significant source of drag is actually derived from the high pressure under the wing, which causes air to flow up over the wingtip and spin off in a vortex. These vortices produce what is called induced drag and are powerful enough to disrupt aircraft flying too closely to one another—one reason for the carefully monitored spacing between flights at takeoff and in the air. Induced drag hampers aircraft performance, cutting into fuel mileage, range, and speed.

In 1897, British engineer Frederick W. Lanchester conceptualized wing end-plates to reduce the impact of wingtip vortices, but modern commercial technology for this purpose traces its roots to pioneering NASA research in the 1970s. At the time, NASA's Aircraft Energy Efficiency (ACEE) program sought ways to conserve energy in aviation in response to the 1973 oil crisis. As part of the ACEE effort, Langley Research Center aeronautical engineer Richard Whitcomb conducted computer and wind tunnel tests to explore his hypothesis that a precisely designed, vertical wingtip device—which Whitcomb called a "winglet"—could weaken wingtip vortices and thus diminish induced drag. Less drag would translate into less fuel burn and better cruise efficiency. The winglet concept provided a better option than simple wing extensions which, while offering similar aerodynamic benefits, would require weight-adding strengthening of the wings and could render a plane too wide for airport gates.

After evaluating a range of winglet designs, Whitcomb published his findings in 1976, predicting that winglets employed on transport-size aircraft could diminish induced drag by approximately 20 percent and improve the overall aircraft lift-drag ratio by 6 to 9 percent.

Whitcomb's research generated interest in civil and military aviation communities, leading to flight testing that would not only confirm his predictions, but help popularize the winglet technology now found on airplanes around the world.

Partnership

In 1977, NASA, the U.S. Air Force, and The Boeing Company, headquartered in Chicago, initiated a winglet flight test program at Dryden Flight Research Center. Whitcomb's Langley team provided the design, and Boeing, under contract with NASA, manufactured a pair of 9-foot-high winglets for the KC-135 test aircraft provided by the Air Force.

Whitcomb was validated: The tests demonstrated a 7-percent increase in lift-drag ratio with a 20-percent decrease in induced drag—directly in line with the Langley engineer's original findings. Furthermore, the winglets had no adverse impact on the airplane's handling. The Dryden test program results indicated to the entire aviation industry that winglets were a technology well worth its attention.

The 1970s were an important decade for winglet development for smaller jet aircraft, with manufacturers Learjet and Gulfstream testing and applying the technology. Winglets for large airliners began to appear later; in 1989, Boeing introduced its winglet-enhanced 747-400 aircraft, and in 1990 the winglet-equipped McDonnell Douglas

During the 1970s, the focus at Dryden Flight Research Center shifted from high-speed and high-altitude flight to incremental improvements in technology and aircraft efficiency. One manifestation of this trend occurred in the winglet flight research carried out on this KC-135 during 1979 and 1980.

Aviation Partners Boeing manufactures and retrofits Blended Winglets for commercial airliners. The technology typically produces a 4- to 6-percent fuel savings, which can translate to thousands of gallons of fuel saved per plane, per year.

MD-11 began commercial flights following winglet testing by the company under the ACEE program.

In 1999, Aviation Partners Boeing (APB) was formed, a partnership with Seattle-based Aviation Partners Inc. and The Boeing Company. The companies created APB initially to equip Boeing Business Jets, a 737 derivative, with Aviation Partners' unique take on the NASA-proven winglet technology: Blended Winglets.

Product Outcome

Like other winglet designs, APB's Blended Winglet reduces drag and takes advantage of the energy from wingtip vortices, actually generating additional forward thrust like a sailboat tacking upwind. Unlike other winglets that are shaped like a fold, this design merges with the wing in a smooth, upturned curve. This blended transition solves a key problem with more angular winglet designs, says Mike Stowell, APB's executive vice president and chief technical officer.

"There is an aerodynamic phenomena called interference drag that occurs when two lifting surfaces intersect. It creates separation of the airflow, and this gradual blend is one way to take care of that problem," he says.

APB's Blended Winglets are now featured on thousands of Boeing aircraft in service for numerous American and international airlines. Major discount carriers like Southwest Airlines and Europe's Ryanair take advantage of the fuel economy winglets afford. Employing APB's Blended Winglets, a typical Southwest Boeing 737-700 airplane saves about 100,000 gallons of fuel each year. The technology in general offers between 4- and 6-percent fuel savings, says Stowell.

"Fuel is a huge direct operating cost for airlines," he explains. "Environmental factors are also becoming significant. If you burn less fuel, your emissions will go down as well." APB winglets provide up to a 6-percent reduction in carbon dioxide emissions and an 8-percent reduction in nitrogen oxide, an atmospheric pollutant. The benefits of winglets do not stop there, Stowell explains. Reduced drag means aircraft can operate over a greater range and carry more payload. Winglet-equipped airplanes are able to climb with less drag at takeoff, a key improvement for flights leaving from high-altitude, high-temperature airports like Denver or Mexico City. Winglets also help planes operate more quietly, reducing the noise footprint by 6.5 percent.

If all the single-digit percentages of savings seem insignificant on their own, they add up. In 2010, APB announced its Blended Winglet technology has saved 2 billion gallons of jet fuel worldwide. This represents a monetary savings of $4 billion and an equivalent reduction of almost 21.5 million tons in carbon dioxide emissions. APB predicts total fuel savings greater than 5 billion gallons by 2014.

APB, the only company to currently both manufacture and retrofit winglets for commercial airliners, is currently equipping Boeing vehicles at the rate of over 400 aircraft per year. It is also continually examining ways to advance winglet technology, including spiroid winglets, a looped winglet design Aviation Partners first developed and successfully tested in the 1990s. That design reduced fuel consumption more than 10 percent.

While winglets require careful customization for each type of plane, they provide effective benefits for any make and model of aircraft—even unmanned aerial vehicles. Consider other winglet designs on commercial carriers, as well as blended and other winglets on smaller jets and general aviation aircraft, and the impact of the original NASA research takes on even greater significance.

"Those flight tests put winglets on the map," says Stowell. ❖

Blended Winglets™ is a trademark of Aviation Partners Inc.

Sensor Systems Collect Critical Aerodynamics Data

Originating Technology/NASA Contribution

The next time you blow out a candle, watch how the smoke behaves. You will see that it rises first in an even stream. At a certain point, that stream begins to break up into swirls and eddies as the smoke disperses.

Air flowing over a wing demonstrates similar behavior. The smooth, even movement of air over a wing is called laminar flow and is what allows aircraft to fly efficiently. Airflow that separates from the wing's surface and breaks up into unsteady vortices is called turbulence, which is responsible not only for those moments of bumpy discomfort during flights but also increased drag, which reduces efficiency and performance. Understanding and controlling the influence of these aerodynamic forces on a wing can lead to aircraft that fly more safely, use less fuel, and carry greater payloads.

In 1981, aerodynamics researcher Siva Mangalam joined NASA's Langley Research Center to work with Werner Pfenninger, a renowned expert in the field of laminar flow control. Mangalam collaborated with Pfenninger on designing a series of unmanned aerial vehicle (UAV) airfoils, one of which has since been incorporated into the Global Hawk UAV, which this year began a series of NASA scientific missions. Through the course of his design work, Mangalam became interested in a basic deficiency in aerodynamics research: While there were extensive systems for wind tunnel testing, there were no quantifiable sensing systems for measuring aerodynamic forces and moments during actual flight.

Mangalam explains that, when an aircraft is in flight, there are three sets of forces in play apart from propulsion: aerodynamics, structural oscillations of the airframe, and forces created by actuators like flaps or tail surfaces. Currently, Mangalam says, the impact of aerodynamic forces on an aircraft is understood only after the fact; turbulence, for example, is only detected after the aircraft's structure responds to it.

Dryden Flight Research Center's F-15B test bed aircraft flew several flights to determine the location of sonic shockwave development as air passes over an airfoil. The shock location sensor (above), developed by Tao of Systems Integration Inc., isolated the location of the shock wave to within a half-inch.

"Today we measure the impact of aerodynamic forces primarily through structural measurement devices such as accelerometers, gyroscopes, and strain gauges. We need sensing systems to measure aerodynamics independently of structures and actuation," says Mangalam. An aircraft's wings, for example, act like low-pass filters: They do not respond quickly enough to aerodynamic disturbances like turbulence to allow structural sensors to capture what the forces are doing in real time.

"Turbulence can be very fast acting, while the structure is slow acting," Mangalam says. This means that not only do researchers not know the true aerodynamic conditions influencing an aircraft in flight, but neither do pilots. Without an exact knowledge of the unsteady forces in play, a pilot negotiating sudden turbulence might do the opposite of what is required, Mangalam says. The ability to detect changes in aerodynamic forces before the aircraft's structure responds can lead to better control systems and safer, more efficient, and more comfortable flight.

Partnership

In 1994, Mangalam founded Tao of Systems Integration Inc. (Tao Systems), headquartered in Hampton, Virginia, to develop real-time systems to quantify aerodynamic and structural dynamics for monitoring and improving aircraft performance and safety. With the support of multiple **Small Business Innovation Research (SBIR)** contracts and flight testing opportunities with Dryden Flight Research Center, Tao Systems began crafting sensors and other components with the ultimate goal of developing a first-of-its-kind, closed-loop system to detect, measure, and control aerodynamic forces and moments in flight.

Tao Systems' approach to gathering aerodynamic data is based on a "game-changing concept," Mangalam says. Essential aerodynamic information such as lift and drag, he explains, can be estimated with measurements at two locations: where the airflow attaches to and detaches from the aircraft surface.

The system works based on a kind of reverse windchill factor. By maintaining a constant temperature on an airplane wing, Mangalam says, "we can determine how much heat loss is caused by the airflow. Depending on how much heat is taken, and its distribution, we can capture the critical points."

This simple, inexpensive approach allows the real-time collection of aerodynamic data free of in-flight interferences like electromagnetic or radio frequency interference—a previous barrier to exposed sensors for accurate in-flight aerodynamic measurements.

Product Outcome

Today, Tao Systems has developed three of the four planned components for its closed-loop system. The final component, a controller, is scheduled for completion by 2012. In the meantime, the company's completed components—its SenFlex hot-film sensors, hot-wire and hot-film constant voltage anemometers, and its proprietary signal processing techniques—are providing sensing solutions for a host of research and development applications.

Customers such as Boeing, Lockheed Martin, Northrop Grumman, General Electric, BMW, and Rolls Royce have employed SenFlex sensors for the nonintrusive measurement of airflow and temperature characteristics. Sandia National Laboratories are using the company's technologies to explore ways of improving wind turbine operation.

"Whenever there is a tremendous amount of power available from high winds, these turbines have to shut down because they cannot handle the load," says Mangalam. "If they could have some way of measuring and controlling the loads generated by the gusting winds, the turbines could generate more power and last longer."

Tao Systems' technologies are also employed for examining airflow in buildings to determine ways to save energy by optimizing air conditioning use.

"Wherever there is airflow, the opportunities are plenty," says Mangalam. But he is quick to note the usefulness of his company's technologies underwater as well, for developing and improving underwater turbines generating electricity from waves and tidal forces. Using its SBIR-developed components, Tao Systems is also working with the U.S. Navy to create a control system for an underwater propeller blade that can alter its shape to maintain high performance in a range of hydrodynamic conditions.

While Tao Systems' customers continue to make use of its innovations, the company has its eye on the potential benefits its completed closed-loop flight control system will offer. Mangalam says that once the system is available, engineers will be able to use it to develop advanced adaptive control systems and flexible wing structures, allowing an aircraft to respond to—and even take advantage of—aerodynamic conditions as they happen, improving flight safety and comfort, increasing fuel efficiency and payload capacity, and extending aircraft lifespan by reducing structural strain and fatigue. Advancements like this would provide human aviation with capabilities closer to those of nature's perfect aerodynamic creation.

"We could fly by feel," he says, "like a bird." ❖

SenFlex® is a registered trademark of Tao of Systems Integration Inc.

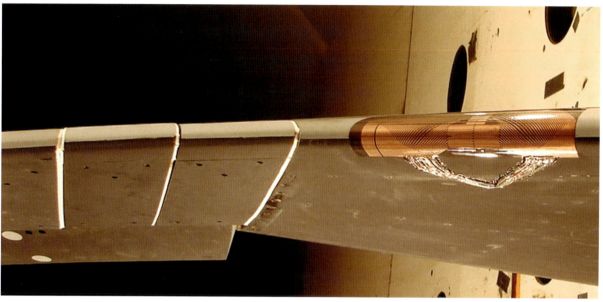

The Tao Systems sensor array, seen here on the leading edge of a Northrop Grumman SensorCraft flexible wing model, was used to successfully demonstrate gust load alleviation. Tao Systems' sensors could one day enable flexible wing aircraft that respond in real time to aerodynamic conditions, resulting in bird-like flight.

Coatings Extend Life of Engines and Infrastructure

Originating Technology/NASA Contribution

Every time a jet engine is started, it goes through a thermal cycle of extreme temperatures, reaching as high as 2,700 °F within the engine's combustor. Over time, the expansion and contraction of engine parts caused by this cycle lead to cracking and degradation that shortens an engine's lifespan and eventually necessitates costly replacement.

Among the many ways that NASA works to advance aviation are efforts to improve the life and performance of jet, or gas turbine, engines. Glenn Research Center scientist Dongming Zhu performed groundbreaking work developing thermal barrier coatings (TBCs) that outperform other TBC technologies, providing an unrivaled means of protecting these engines from the degrading effects of thermal stresses, prolonging their lifespans, and enhancing their reliability and fuel efficiency.

TBCs are ceramic coatings with low thermal conductivity, insulating the metal parts they are applied to and reducing thermal fatigue. The ceramic component is typically composed of zirconium oxide (zirconia) stabilized in a particular crystal structure through the addition of yttrium oxide (yttria). Through the addition of other oxides, Zhu created modified yttria-stabilized zirconia TBCs with both initial and post-exposure thermal conductivities that are even lower than existing coatings. Zhu's breakthrough thermal and environmental barrier coating work was recognized by *R&D Magazine* in 2007 with an "R&D 100" award as one of the year's 100 most technologically noteworthy inventions.

Now, through collaboration with a private industry coatings leader, Zhu's NASA research is helping extend the service of military aircraft.

Partnership

"NASA is the world leader in thermal barrier coatings," says Andrew Sherman, founder and CEO of MesoCoat Inc., based in Euclid, Ohio. A subsidiary of Powdermet Inc., also based in Euclid and itself a NASA partner through the **Small Business Innovation Research (SBIR)** program, MesoCoat was founded in 2007 to develop and commercialize new coating techniques. Through a collaborative agreement with Glenn and under contract with the U.S. Air Force, MesoCoat is employing a specific composition of Zhu's TBC technology to prolong the lifespan of engines in the Air Force's aging, legacy aircraft. Since the commercial application is currently only for government use, no licensing agreement was required, though MesoCoat is engaged in discussions for commercial licensing terms.

"It's very difficult to get new parts for these older engines," says Sherman, "so the Air Force needs new technologies to extend the life of components" such as combustor cans and afterburner nozzles. The company's NASA-derived technology is providing an ideal, cost-effective solution for this need.

The ZComP 844 nanocomposite thermal barrier coating (TBC) has about half of the thermal conductivity of standard thermal barriers. Applying the TBC to engine components can result in a 50-percent increase in component life as a result of reduced thermal stress.

Product Outcome

Branded ZComP 844, MesoCoat's nanocomposite TBC has cluster formations that inhibit radiation transfer in the TBC and improves the coating's stability, Sherman explains, so that it has about half of the thermal conductivity of conventional thermal barriers.

"The NASA solution allows us to reduce the thermal conductivity, which reduces the engine part temperature as well as smoothes out thermal cycles, making them slower, much more uniform, and taking a lot of the thermal stresses off the part," says Sherman. "By halving the thermal conductivity, we're shaving around 30 to 50 degrees off the part temperature. As a result, you're looking at a 50-percent increase in component life." Reducing the thermal stresses on engine components also results in better performance and fuel efficiency.

Other methods for combating thermal fatigue, such as applying thicker layers of other TBCs or changing engine operating conditions, are either less effective or entirely unfeasible, Sherman says. The ZComP 844 TBC can be applied at almost no additional cost, he explains, by simply inserting the enhanced TBCs into the normal maintenance cycle when the coatings are stripped and replaced.

While MesoCoat's NASA-developed TBC promises to provide longer component life, improved fuel economy, and better operating efficiency for other commercial engine applications in the future, the company is engaged in another NASA partnership that may soon provide similar lifespan-enhancing benefits to the Nation's aging infrastructure.

Combating the effects of corrosion—a significant threat to highways, bridges, and other structures around the country—costs the U.S. economy over $270 billion annually, according to the Federal Highway Administration. MesoCoat has developed a complete metal cladding system, called CermaClad, that provides a quickly applicable, environmentally friendly, and cost effective method for extending the maintenance-free life of steel structures like bridge beams, water and oil pipes, and rebar. Employing inexpensive, inert nanocomposite cermet (ceramic and metallic) materials that are bonded to the metal surface using a high-density infrared (HDIR) arc lamp, CermaClad provides a cheaper, lighter, and nontoxic alternative to welded cladding methods and bioactive, carcinogenic materials like chromates. The resulting coating, which can be applied at rates 100 times faster than weld overlays, can prevent any degradation for periods up to 100 years, Sherman says.

While the technology is not yet commercial, MesoCoat has validated the product with "a number of very large customers" and is creating demonstration scale prototypes and fusion clad components thanks to a Space Act Agreement with Glenn that allows the company to make use of the Center's 200-kilowatt HDIR lamp system.

The arc lamp technology was originally used by NASA for simulating heat fluxes that occur during the reentry of spacecraft into Earth's atmosphere, as well as for thermal testing of combuster liners. As a student at the University of Cincinnati and a NASA intern, Craig Blue adapted the technology for coatings applications and is currently advancing its uses at the Oak Ridge National Laboratory (ORNL) in Oak Ridge, Tennessee. MesoCoat secured the commercialization rights to the "R&D 100" award-winning innovation from ORNL and expects CermaClad to provide a solution that overcomes the biggest obstacle to infrastructure improvement: cost.

"It's pretty much understood that we could solve the problem of deteriorating infrastructure if we wanted to. It's a sheer economic issue," Sherman says. "The real novelty of CermaClad is the productivity of it, which allows us to apply metal and ceramic coatings at the same rate you can spray paint today."

A key to this potential becoming reality, he says, is MesoCoat's NASA partnership.

"Bringing NASA expertise and facilities to bear allows us to outcompete anyone else in the world." ❖

ZComP 844™ and CermaClad™ are trademarks of MesoCoat Inc.

MesoCoat Inc.'s CermaClad technology uses a high-density infrared arc lamp (above) to bond the cladding material to metal surfaces, preventing corrosion for up to 100 years. While this technology extends the lifespan of infrastructure like steel beams and pipes, the company's ZComP 844 TBC extends the life of gas turbine engine components (right) and hardly adds any cost to normal maintenance cycles.

Public Safety

NASA's emphasis on safety translates not just to its rocket launches and laboratory practices, but also to our everyday lives. The technologies featured in this section:

- Optimize Local Weather Prediction
- Eliminate Icing Danger for UAVs
- Rescue Entire Planes
- Advance UAVs for Science, Military
- Support Emergency Communication
- Assess Structural Health
- Detect Explosives and Chemical Agents
- Advance Imaging for Security, Industry

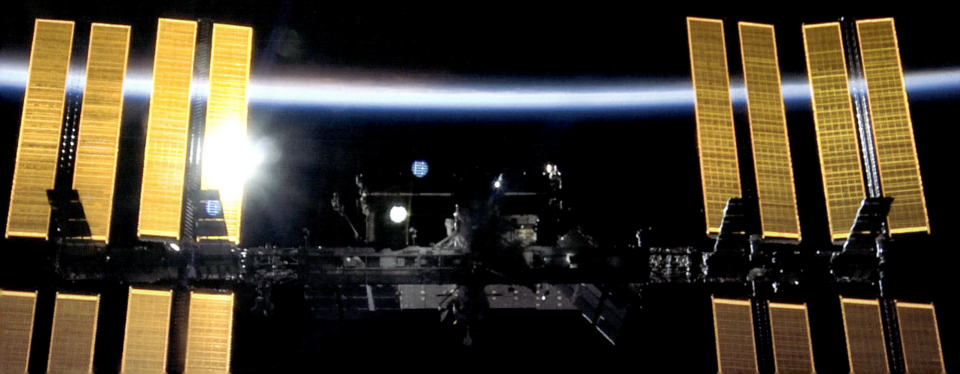

Radiometers Optimize Local Weather Prediction

Originating Technology/NASA Contribution

One of the greatest dangers to aircraft—playing a role in numerous destructive and fatal accidents around the world—comes in the form of droplets of water. Clouds are made up of tiny water particles with diameters typically between 10 and 50 microns. In clean air, cloud droplets can exist in liquid form down to temperatures as low as -40 °C. These subfreezing, liquid clouds are referred to as being "supercooled." As soon as supercooled droplets contact an aircraft ascending or descending through the cloud cover, they form layers of ice on any unprotected surface, including the leading edges of wings and rotor blades, tails, antennas, and within jet engines. This ice accretion can cause engine damage and dramatically affect the aerodynamics of the aircraft. (On the leading edge of a wing, an ice layer about as thick and rough as a piece of coarse sandpaper can be responsible for as much as a 30-percent decrease in lift and a 40-percent increase in drag.) This can lead to reduced performance and even catastrophic loss of control.

As part of its aeronautics research, NASA has extensively investigated the icing problem, leading to numerous spinoff technologies that are helping reduce the threat. Glenn Research Center has led the Agency's efforts, testing thermal, chemical, and mechanical anti-icing technologies in its Icing Research Tunnel; developing software tools for modeling ice growth and the impact of icing on aircraft performance; and producing pilot training aids for flight in icing conditions.

Partnership

One way of mitigating the dangers of ice buildup is through the accurate, real-time identification of icing conditions, and researchers at Glenn have studied ways to detect supercooled water droplets in the flight paths of aircraft in and out of airports. One such method involves combining weather radars with devices called microwave

The NASA Narrow-beam Multi-waveband Scanning Radiometer (NNMSR), seen here with the CHILL radar in the background, can determine possible icing conditions along a flight path.

radiometers, which measure the energy emission of liquid water and water vapor in the atmosphere at microwave frequencies (between 1–1,000 gigahertz). Combining the ability of the radar to detect cloud and hydrometeor particles—particles big enough to fall, like rain and hail—with the radiometer's ability to detect liquid and vapor levels, provides a comprehensive picture of particle size, type, and distribution within clouds—essential information for determining icing risk.

Glenn partnered with Radiometrics Corporation, of Boulder, Colorado, to advance microwave radiometer technology for the detection of icing conditions. Supported by Phase I and II **Small Business Innovation Research (SBIR)** contracts, Radiometrics identified distinct, polarized signatures for liquid and ice cloud particles. These findings instigated further investigation with a narrower beam radiometer, which the company invented through additional Phase I and II SBIR agreements with Glenn. The resulting technology, a pencil-beam radiometer called the NASA Narrow-beam Multi-waveband Scanning Radiometer (NNMSR), is "a pioneering instrument that is seeing things in nature that have never been seen before," says Randolph Ware, vice president of sales and marketing for Radiometrics. "You can locate this instrument in an airport, look at a narrow beam width along a flight path, and detect the supercooled liquid that creates the icing hazard."

Testing in conjunction with Colorado State University's CHILL radar (named after its original location in CHicago, ILLinois) supported the NNMSR's ability to detect icing conditions, and further evaluation will take place at Glenn. The development of the NNMSR, Ware says, is a credit to the SBIR program.

"SBIR funding is how we continue to create new technologies," he says. "It is one of the most effective ways for the government to stimulate innovation. Small businesses are the place where jobs grow in this country, so this is a very powerful and important program."

Product Outcome

While a commercial version of the NASA radiometer is still in the works, Radiometrics has brought to market a modular radiometer, originally developed under the NASA SBIR to enable the pencil-beam technology. The modular radiometer represents the company's fourth generation of radiometer products.

"These are sealed radiometers that are actually submersible and have much better performance in many ways," says Ware.

Radiometrics Corporation's modular polarized radiometers (left) are used in soil, snow, ocean, and atmospheric research. The company's other radiometers (below) are powerful tools for short-term weather forecasting.

Radiometrics' radiometers provide thermodynamic profiling of atmospheric conditions, offering continuous temperature, humidity, and liquid profiles up to 10 km in height. These are parameters, Ware says, that are invisible to the eye but define local weather, making the devices powerful tools for improving local, short-term weather forecasting and producing high-accuracy nowcasting (the forecasting of weather within a 6-hour timeframe). The radiometers function robustly in the presence of radio interference and in all weather conditions, and are fully operable via the Internet from anywhere in the world. Essentially, they are hyperspectral atmospheric observatories that can see in all directions, providing complementary measurements to those gathered by standard weather radiosondes.

The company's customers employ its products for the prediction of weather conditions like fog and convective storms, which are known to produce hail, strong winds, flash floods, and tornadoes. The radiometers are also used to detect ideal situations for weather modification; by locating high concentrations of the same supercooled droplets that create icing hazards, the instruments can indicate prime cloud targets for seeding with nuclei particles like silver iodide or dry ice.

Chinese officials relied on a network of Radiometrics' radiometers for anticipating short-term weather and weather modification needs for the Beijing 2008 Olympic Games, and a Radiometrics device was also situated in Whistler, Canada, for the same purposes during the 2010 Winter Olympics. Ware says that the company's networks are going up in China, India, Japan, Korea, and Europe, and are set to be included in a U.S. profiler network established by the National Weather Service. A Radiometrics radiometer is also part of an aviation weather decision support system established at Dubai International Airport to predict short-term weather for air traffic control.

Ware also notes that the company's modular, polarimetric radiometers have applications aboard ships and for coastal observatories, looking down at the sea surface to detect salinity and temperature. (Polarized radiometers have the same advantage as polarized sunglasses used by fishermen—they are able to see past the reflected surface glare to indentify underlying signals.)

The modular radiometers are also being used in soil moisture studies; to look at snow pack age, depth, and water content; and to assist with satellite sensor calibration and satellite link loss and prediction. The U.S. Department of Energy has ordered a number of Radiometrics' fourth-generation, modular radiometers for use in the Atmospheric Radiation Monitoring program, a significant contributor to global climate change research. The same qualities that make a radar/radiometer combination ideal for aircraft icing conditions detection research, says Ware, make the NASA SBIR-derived instrument a key tool for understanding the transfer of radiation through clouds, which is dependent on the distribution of particles in clouds and is one of the biggest uncertainties in global climate models.

In keeping with the typical cycle of successful technology transfer, NASA is also benefiting from Radiometrics' commercialized, SBIR-derived technology. NASA's Jet Propulsion Laboratory has purchased modular radiometers for its Deep Space Network, a worldwide antenna network supporting solar system exploration. ❖

Energy-Efficient Systems Eliminate Icing Danger for UAVs

Originating Technology/NASA Contribution

One of the more remarkable developments in aviation in recent years has been the increasing deployment of unmanned aerial vehicles, or UAVs. Since the invention of the first UAV in 1916, these remotely—or sometimes autonomously—controlled vehicles have become invaluable tools for military reconnaissance and combat, cargo transport, search and rescue, scientific research, and wildfire monitoring. Free from having to accommodate the safety needs and endurance limits of a pilot, UAVs are capable of flying extended missions and venturing into hazardous and remote locations.

There is one common aviation danger, however, that plagues unmanned and manned aircraft alike. In certain atmospheric conditions, layers of ice can build up on an aircraft's leading surfaces, dramatically affecting its aerodynamics and resulting in decreased performance or even complete loss of control. Lightweight UAVs are particularly susceptible to icing problems, and the potential damage icing conditions can cause to these expensive vehicles can render their operation unfeasible in certain weather. This is particularly troubling for military UAV applications, in which icing conditions can lead to aborted missions and the loss of crucial tactical capabilities.

Countering the threat of icing has been one of NASA's main aeronautics goals. NASA research has led to a variety of deicing technologies that are now making aviation safer for all aircraft. One such solution, invented by Ames Research Center engineer Leonard Haslim, employs a pair of conductors embedded in a flexible material and bonded to the aircraft's frame—on the leading edge of a wing, for example. A pulsing current of electricity sent through the conductors creates opposing magnetic fields, driving the conductors apart only a fraction of an inch but with the power to shatter any ice buildup on the airframe surface into harmless particles. Haslim called his invention an electroexpulsive separation system (EESS), or the "ice zapper," and it earned him NASA's "Inventor of the Year" award in 1988. Decades later, this technology is now proving to be an ideal solution to the UAV icing dilemma.

Partnership

Aeronautics engineer Mark Bridgeford was wandering through a technology trade show when he came across a booth where a pair of representatives from Ames was displaying NASA innovations, including something that attracted Bridgeford's attention. The representatives were setting poker chips on a device and popping them into the air to demonstrate its capabilities; it was a model of Haslim's invention. Bridgeford immediately realized

NASA's Altair UAV was developed by General Atomics as a long-endurance, high-altitude platform for development of UAV technologies and environmental science missions. IMS-ESS deicing systems are providing icing protection for other General Atomics UAVs for military applications.

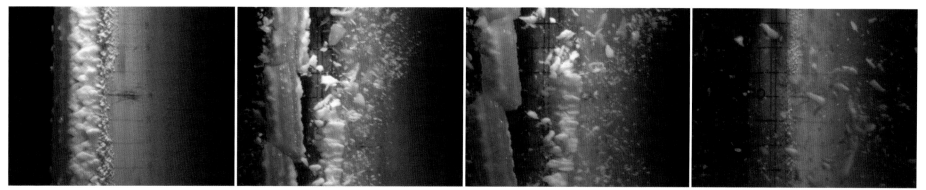
An electric current drives apart a pair of conductors embedded in the leading edge of a wing, shattering any ice buildup on the surface into harmless particles.

that there was nothing like the EESS on the market and learned that Ames was actively seeking to transfer the technology to the public sector.

"I recognized from the beginning that this was a home-run technology," Bridgeford says. In 1995, he and business partner Richard Olson formed Temecula, California-based Ice Management Systems (IMS, now known as IMS-ESS after merging with cable and harness assembly manufacturers Electro-Support Systems in 2007) and licensed the patent for Haslim's innovation from Ames that same year.

Through the course of extensive research and development, IMS built upon the existing NASA concept, creating an energy-efficient power system and a composite, leading-edge cuff with the conductive actuators embedded internally in the carbon-fiber structure. The company conducted extensive testing in icing tunnels, including the Icing Research Tunnel at Glenn Research Center, proving the technology's consistent effectiveness in removing ice from airframe surfaces.

Product Outcome

IMS-ESS was first featured in *Spinoff* 2001, following the company's sale of several of its electroexpulsive deicing systems (EEDS) to Thompson Ramo Wooldridge for use on the TRW (now the Northrop Grumman) Hunter UAV. Around the same time, Bridgeford says, major UAV manufacturer General Atomics, producer of the well-known military Predator UAVs, was beginning work on its Sky Warrior UAV program for the U.S. Army. Well aware of the icing problems that affected the usefulness of its UAVs, General Atomics brought IMS-ESS into its development program for the Warrior. IMS-ESS' NASA-derived deicing systems are currently in production for use on the Warrior, just one of the company's successes in the field of UAV deicing solutions.

"One reason we have hit a spot with the UAV business is the simple fact that our system utilizes so little energy," he says. The system uses around 600–900 watts—"unheard of for ice protection"—whereas typical thermal deicing systems use tens of thousands of watts, requiring a large generator impractical for use on a lightweight UAV.

Other solutions are equally problematic, Bridgeford says. "Weeping wing" deicing systems, which coat wing surfaces with anti-ice agent ethylene glycol, require the UAV to carry a weighty onboard supply of the chemical that may run out over the course of a long-duration flight. Pneumatic boot systems, which inflate to break away ice buildup, require a certain amount of ice accumulation before they can operate successfully. IMS-ESS systems, in contrast, are lightweight, low power, and are effective for any ice thickness.

"With the electroexpulsive technology, UAVs are able to incorporate continuous, year-round ice protection into their airframes," Bridgeford says.

IMS-ESS is taking a big step forward this year, Bridgeford notes. Besides the Sky Warrior, the company's deicing systems are going into production for use on the Thales UK Watchkeeper Tactical UAV for the British Army, and Bridgeford expects other UAV programs to follow suit in the near future. IMS-ESS is also set to pursue Federal Aviation Administration certification for its systems, a major step toward delivering the benefits of the company's NASA-derived deicing system to piloted aircraft.

While Bridgeford acknowledges the challenges facing small businesses when bringing sophisticated technologies to fruition, he sees NASA as a source of unique technology for commercial applications.

"This company was launched on NASA technology," he says. "I would like to see more of this kind of exchange." ❖

Sky Warrior® is a registered trademark of General Atomics Aeronautical Systems Inc.

Rocket-Powered Parachutes Rescue Entire Planes

Originating Technology/NASA Contribution

When Boris Popov was 8 years old, he took one of his mother's sheets and some thread, made a parachute, climbed a tree, and jumped. The homemade chute did little to break Popov's fall; his father took the disappointed boy aside and said, "Son, you've got to start higher."

Years later in the mid-1970s, recent college graduate Popov was hang gliding over a lake when the boat that was towing him accelerated too quickly, ripping the control bar from his hands. Some 500 feet in the air, Popov's glider went into a spiral, coming apart as Popov plummeted to the water. As he fell, Popov realized that if he only had some kind of parachute, he could have been saved. Before impact, he promised himself that, if he survived, he would create a solution that would save people in these types of emergency situations.

Decades later, the U.S. air transportation system was suffering its own kind of free fall. The terrorist attacks of 9/11 led to stringent security measures that complicated and slowed down air travel. Even as the industry recovered from the effects of the attacks, increased flights and passenger demand strained the National Airspace System (NAS) at levels never before experienced. At the same time, NASA was exploring ways of extending aviation to rural America using smaller general aviation (GA) aircraft and local community airports. The NASA Small Aircraft Transportation System (SATS) project envisioned an on-demand, point-to-point, widely distributed transportation system relying on small aircraft (4-10 passengers) operating out of the Nation's more than 5,400 public-use landing facilities. With about 98 percent of the population living within 20 miles of at least one such airport, SATS could provide cheaper, faster, and more practical options for business and leisure travel, medical services, and package delivery.

> *"BRS is a classic example of taxpayers' money being spent on research that has translated into 246 lives saved."*

Though the SATS project concluded its research in 2006, the pursuit of a nationwide GA transportation system continues through other initiatives like NASA's Green Flight Centennial Challenge, scheduled for 2011, which encourages competing teams to maximize fuel efficiency for personal aircraft, as well as reduce noise and improve safety. Technological advances are still necessary, however, to make such a system viable, such as improving the safety of small aircraft. One solution has come in the form of an invention developed by Popov, who having survived his fall, began investigating methods of ballistically deploying parachutes for aircraft in emergency situations. Today, with the help of a NASA partnership, the parachute that Popov wished for when plunging to Earth is saving hundreds of small aircraft pilots from a similar fate.

Partnership

Popov founded Ballistic Recovery Systems Inc. (now BRS Aerospace) of Saint Paul, Minnesota, in 1980. He formed the company to commercialize his solution to personal aircraft accidents like the one he experienced: a whole aircraft parachute recovery system. Soon BRS was developing parachutes for hang gliders, ultralights, and experimental aircraft, and the company received Federal Aviation Administration certification for a retrofit system for the Cessna 150 GA airplane. The company's innovative safety solution for small aircraft led to **Small Business Innovation Research (SBIR)** contracts with Langley Research Center aimed at advancing the BRS parachute system for use with larger and heavier GA aircraft. The NASA funding helped BRS with the development of thin-film parachutes, continuous reinforcement manufacturing methods that result in stronger parachutes, and smart deployment devices—all of which help overcome one of the main obstacles to whole-aircraft parachute systems for larger vehicles: reducing bulk and weight while maintaining parachute strength.

"You can't have a 50-gallon drum full of parachute in the back of a Cessna. It's not going to work," Popov says. Just as important as the research and development funding for BRS, he says, was NASA's support of its parachute system.

"One of our primary needs for working with NASA was to promote and encourage the concept of a ballistic parachute on aircraft," Popov says. "There was a lot of skepticism that this system could even work. NASA was very proactive in creating a safety mentality in general aviation."

This BRS Aerospace Inc. parachute system, designed for sport aircraft, deploys its chute (contained in the white canister) in less than 1 second, thanks to a solid rocket motor (the black tube on top).

With the help of NASA funding, BRS developed parachutes that have saved hundreds of small aircraft—and their pilots and passengers. Here, a Cirrus SR20's parachute deploys at over 100 miles per hour, arresting the plane's descent. BRS parachute systems are standard equipment on Cirrus aircraft.

Product Outcome

The BRS parachute system—first featured in *Spinoff* 2002—is deployed by a solid rocket motor activated when the pilot pulls on the cockpit handle release. The rocket fires at over 100 miles per hour and extracts the parachute in less than 1 second. Thanks to a patented shock attenuation device, the chute opens according to the speed of the aircraft; at high speeds, the chute opens only 25 percent for the first few seconds to reduce airspeed to the point where the chute can open fully and still sustain the opening shock. (The lightweight parachute material has to sustain the force of the rocket deployment, as well as the force of the aircraft.) At low speeds and altitudes, the chute opens quickly and completely to ensure rescue.

The system's versatility makes it effective in a range of accident situations, from mid-air collisions and structural failure to a spiral dive or stall spin. The parachute arrests the descent of the entire aircraft and deposits the vehicle and its occupants on the ground with a typical impact force equivalent to falling 7 feet, which is largely absorbed by the aircraft's landing gear and seats. Not only are lives saved, but in many incidents, expensive aircraft are preserved to fly again.

BRS has sold more than 30,000 systems worldwide since its founding. The parachute is now standard equipment on the Cirrus SR20 and SR22 planes, the Flight Design CT light-sport aircraft (LSA), the Piper Aircraft PiperSport LSA, and as an option on the new Cessna 162 Skycatcher. The company is projecting sales of close to $20 million this year.

"Our system is standard equipment on the world's top selling single-engine aircraft, Cirrus. It's standard equipment on the world's top selling LSA, the CT. The number one producer of ultralights has our product as standard equipment. You can see a trend here," Popov says.

BRS also produces parachute systems for military unmanned aerial vehicles, military cargo parachutes, and military training aircraft recovery parachutes. On training aircraft, if the pilot has to eject, "you basically have a 5,000-pound bomb that could go unpiloted down into a neighborhood," Popov says. "We, however, can bring down the pilot and trainer aircraft safely to the ground."

While parachutes for larger aircraft are still in the works, BRS does have a system designed for small jets, and its NASA partnership has provided the company with the technology that may eventually enable parachutes for commercial airlines and jets. In the meantime, Popov welcomes the role NASA has played in helping turn the promise he made to himself that day at the lake into a reality for the 246 people whose lives have been saved by the BRS parachute so far.

"BRS is a classic example of taxpayers' money being spent on research that has translated into 246 lives saved," he says. "That's a justifiable and profound benefit."

He tells a favorite story about a grandfather flying a Cirrus SR20 over the Canadian Rockies with his grandkids in the back seat. The grandfather lost control of the plane, which became inverted at night in the mountains. "You're likely not going to recover from that," Popov says. The grandfather deployed the parachute, and the plane settled gently on the side of mountain, where a rescue helicopter found it the next day. After being hoisted out by a helicopter and flown to a nearby airstrip, they put on a new prop and landing gear and flew the plane out.

"This grandfather thought he may have just killed himself and his grandkids, but when he pulled the handle and felt the parachute deploy, he knew he had just prevented that from happening," Popov says.

"How many millions of dollars is that worth?" ❖

Technologies Advance UAVs for Science, Military

Originating Technology/NASA Contribution

Greek mythology tells of the inventor Daedalus using wings of his own fashioning to escape from imprisonment on the island of Crete. In 1988, a similar adventure was launched, though in this case carbon-fiber composites, gears, and driveshafts featured instead of wax and feathers.

A year earlier, a group of students, alumni, and professors from the Massachusetts Institute of Technology (MIT) gathered at Dryden Flight Research Center to begin a series of test flights for what they hoped would be a record-setting effort. Inspired by the Greek myth, the team built and tested three lightweight, human-powered aircraft designed to reenact Daedalus' (according to the tale) 115-kilometer flight. After numerous test flights of the three aircraft (and one crash), the 69-pound Daedalus 88 launched from Crete in April 1988. Powered only by the pedaling of the pilot, a Greek champion cyclist, the aircraft flew nearly 4 hours and 199 kilometers before winds drove it into the sea just off the coast of the island of Santorini. (If this calls to mind the demise of Daedalus' son Icarus, do not worry; the pilot swam to shore.)

Setting distance and duration records for human-powered flight that are still unmatched today, the Daedalus project provided NASA and the MIT team the opportunity to explore new technologies for lightweight aircraft and high-altitude, long-duration flight. Also from this effort came the kernel of a company that—with the help of NASA partnerships—is producing some of the world's most advanced aviation technologies.

Partnership

In 1989, John Langford founded Aurora Flight Sciences Corporation in a small office in Alexandria, Virginia. Langford had managed the Daedalus project and saw great potential in applying the technologies developed for that effort to the innovation of high-altitude unmanned aerial vehicles (UAVs) for global climate change research. Almost immediately, Aurora established a pattern of partnership with NASA that continues today.

"NASA has been a critical supporter of Aurora from day one," says Langford. The company has engaged in numerous **Small Business Innovation Research (SBIR)** and **Small Business Technology Transfer (STTR)** projects with the Agency, beginning with its initial Ames Research Center SBIR, for the development of a fuel cell-based high-altitude propulsion system, up until its most recent contract in 2009 to create aspirated compressors for a high-altitude, long-endurance (HALE) concept engine for Glenn Research Center. These partnerships have provided opportunities for Aurora on multiple fronts, Langford says.

The Daedalus 88 aircraft is seen here on its last flight at Dryden Flight Research Center in 1988. The aircraft set records for human-powered flight that still hold today.

"There is a technology development function, a personnel development function, and also a collaboration function through SBIRs and STTRs," he explains. "We have bright new talent, and these programs provide a great way for people to explore new ideas."

Aurora, now headquartered in Manassas, Virginia, has also worked with NASA on several unique initiatives. The company developed the Perseus A, Perseus B, and Theseus test bed UAVs for NASA's Environmental Research Aircraft and Sensor Technology (ERAST) program. Designed to encourage the advancement of cost-effective UAVs for HALE science missions, ERAST was instrumental in the creation of new UAV technologies like the Predator B, known as the Altair in its NASA science mission version and as the MQ-9 Reaper for the military. Aurora also designed and created a series of UAVs for potential long-range science missions on Mars.

In addition, a Space Act Agreement with Goddard Space Flight Center and West Virginia University was significant to Aurora's commercial activities today. Through the partnership, Aurora developed low-cost composite materials manufacturing capabilities and opened a manufacturing facility in West Virginia. These outcomes enabled Aurora to provide cost-efficient airframe parts for the Teledyne Ryan (now Northrop Grumman) Global Hawk UAV, designed for the U.S. Air Force.

Product Outcome

Aurora now has 350 employees and has facilities in Mississippi and Massachusetts, in addition to its West Virginia and Virginia operations. The company employs 160 people in its NASA-enabled West Virginia plant, and about one-third of Aurora's work force is dedicated to the company's Global Hawk efforts. Aurora now supplies all the composite structures for Global Hawk, save for the wings.

"This is an example of economic development done right," Langford says. "You want to build up the economy

Through a Space Act Agreement, Aurora Flight Sciences Corporation developed manufacturing techniques that enable the company to provide much of the composite airframe for the Global Hawk UAV (left). These capabilities allowed Aurora to open a manufacturing facility in West Virginia that now employs 160 workers. The company also develops its own cutting-edge UAV designs, such as the GoldenEye 80 ducted-fan aircraft (right) for military surveillance applications.

across the country, and this was a move that NASA participated in that has been very successful."

The partnership has also allowed Aurora to contribute to the use of UAVs for scientific endeavors; NASA's two Global Hawk aircraft began long-duration science missions over the Pacific Ocean in 2010 as part of the Agency's Global Hawk Pacific Mission.

"We are very proud of the fact that parts of those planes were built in our West Virginia facility," Langford says.

Aurora's expertise in advanced aviation development—cultivated in part through its NASA work—has led to the creation of unique robotic aircraft that are providing entirely new takes on UAV design and function. Aurora's GoldenEye and Excalibur aircraft are both vertical takeoff and landing vehicles with science-fiction looks. The GoldenEye 50 and the larger GoldenEye 80 are ducted-fan aircraft designed to provide highly portable surveillance for military applications. Excalibur is an armed tactical UAV that takes off on one side and then rolls over for mission mode, protecting its sensors from any dust kicked up by its propulsion system.

"A lot of aircraft missions are not involved with carrying people, and when you take the people out of the equation, it completely changes the design space," explains Langford. "GoldenEye and Excalibur are examples of designs that have no analog in manned aircraft."

The popularity of UAVs has risen dramatically in recent years, Langford says. "Ten years ago, all the UAVs in the world flew a few thousand flight hours per year, total. This last year, UAVs flew between 500,000 and a million flight hours." Such growth is remarkable, he says, but still hardly compares to the 100 million manned flight hours flown each year worldwide. And science missions are still a vast minority of UAV applications, though Langford predicts that will change with NASA's help.

> "Understanding and protecting our planet is a huge reason why taxpayers should be enthusiastic about NASA."

"We are extremely excited about the renewed emphasis on aeronautics and global change research. It is something only NASA can do, and it fits perfectly with NASA's heritage, mission, and capabilities. Understanding and protecting our planet is a huge reason why taxpayers should be enthusiastic about NASA."

In the meantime, Aurora is continuing work on a number of UAV projects, including its Orion HALL (high-altitude, long-loiter) aircraft, which can fly for 5 days nonstop, and the Centaur optionally piloted airplane. And while the original Daedalus aircraft that inspired Aurora's founding set records with its nearly 4-hour flight, the company is also working on a project that may one day perform flights of positively mythical proportions: The solar-powered Odysseus aircraft may one day be capable of flying for up to 5 years at a time. ❖

Perseus™, Theseus™, GoldenEye™, Excalibur™, and Orion HALL™ are trademarks of Aurora Flight Sciences Corporation.

Inflatable Antennas Support Emergency Communication

Originating Technology/NASA Contribution

Space exploration requires reliable and efficient communication technology. One device currently under development is the inflatable antenna. Due to several unique characteristics—it is lightweight, easy to deploy, inexpensive, and requires low storage volume—inflatable technology is especially well-suited for space applications. Without requiring mechanical actuators or human assembly, something the size of a suitcase could be inflated in space to the size of a basketball court.

In 1960, NASA launched the first of its inflatable space structures in the form of large metalized balloons, the Echo satellites. These were designed to act as communication reflectors to transmit signals from one point on Earth to another. Echo 1A was successfully orbited and was used to redirect telephone, radio, and television signals. Increasingly powerful launch vehicles became available, however, and lightweight inflatable technology was temporarily shelved. In the 1980s and 1990s, interest was renewed due to the cost advantages of the technology, and the space shuttle STS-77 mission successfully deployed an inflatable antenna in space in 1996.

Partnership

With renewed interest brewing in inflatable structures, NASA encouraged further advancements of the technology. ManTech SRS Technologies (formerly SRS Technologies), of Newport Beach, California, received **Small Business Innovation Research (SBIR)** funding from Glenn Research Center in 1997 to develop an inflatable solar concentrator for power generation. The resulting thin polyimide material used to craft the inflatable concentrator was licensed by SRS and commercially produced as a powder, resin, and rolled film (*Spinoff* 1998).

The inflatable antenna can provide backup communication (left) when land lines are down. In the aftermath of Hurricane Ike, GATR Technologies volunteered the use of the antenna to provide high-bandwidth communications for local residents. Soldiers (above) deploy and test an inflatable antenna at Fort Bragg, North Carolina.

It soon became evident that the same basic technology for solar concentrators was applicable for large inflatable antennas, and follow-on SBIR contracts focused on using the polyimide material to develop thin film inflatable antennas for space communication.

Paul Gierow, one of the engineers with SRS at the time, explains, "To make a solar concentrator, you point it at the Sun and focus the energy. The antenna is exactly the same thing, but instead of focusing it on the Sun, you point it at a satellite that is radiating radio frequency (RF) energy. Anything that focuses sunlight energy can typically focus RF energy," says Gierow.

With the help of SBIR funding, SRS modified the concepts and processes for ground-based inflatable antennas. "We came up with an idea to put an antenna in a ball, or sphere. Intuitively you don't think it will work, but it did," says Gierow.

GATR Technologies, of Huntsville, Alabama, was formed in 2004. GATR, which is an acronym for "Ground Antenna Transmit and Receive," licensed the technology from SRS and has provided additional product refinements leading to a ground-based satellite communications system. The company's efforts were enhanced by a U.S. Department of Defense award to mature the ground-based antenna system.

To test the new antenna, including testing of the system's performance for Federal Communication Commission (FCC) certification requirements, GATR entered into a Space Act Agreement with Glenn. As a result, the company gained additional technical understanding of the product and was able to achieve the world's first inflatable antenna certified by the FCC in 2008. This qualified the antenna for a variety of communication applications within NASA, other government agencies, and commercial entities. That same year, SRS and GATR received the "Tibbetts Award" in recognition of small businesses and SBIR

This complete satellite communication system takes about 45 minutes to set up and 15 minutes to take down. When deflated, the flexible antenna rolls up like a sleeping bag and fits in a case with the antenna bag, blower, hoses, and plates. A separate case holds the electronics, including a modem, spectrum analyzer, cables, computer, and power inverter. Each case weighs less than 90 pounds and can be checked as baggage on an airplane.

support organizations exemplifying the types of business, economic, and technical development goals of the SBIR program. *Popular Science* magazine also recognized GATR and designated the inflatable antenna as a "2007 Invention of the Year." In 2010, the system earned recognition as one of the "R&D 100" (a list of the top 100 inventions of the year), granted by *R&D Magazine*. Recently, the company entered into a new Space Act Agreement with Glenn to expand the antenna's frequency range and size, as well as to test and evaluate a larger antenna.

Product Outcome

Since its start in 2004, GATR has sold 60 ground-based inflatable antenna systems. In 2010 alone, the company's goal is to produce 50 more systems. With just under 20 employees, revenue has increased from $3 million in 2008 to $5.6 million in 2009. Most recently, *Inc.* magazine featured GATR on their list of the top 500 fastest growing entrepreneurial companies in the United States.

Capable of providing Internet access, voice over Internet protocol, e-mail, video teleconferencing, broadcast television, and other high-bandwidth communications, GATR's inflatable antenna can be quickly deployed in remote, hard-to-reach areas. The complete system takes about 45 minutes to set up and 15 minutes to take down.

First, the system is anchored to the ground using cables and stakes. Then the four ground mounting plates are oriented toward the targeted satellite at the same time that the azimuth (east or west direction) is set. Using a low-power blower, the antenna then inflates, pointing toward the satellite.

Resembling a large beach ball, the antenna system is made of a flexible reflective fabric that is transparent to RF energy. As airflow to the upper hemisphere increases, it pushes the flexible reflector dish down into a parabolic shape. After that, the ball continues to act as a pressure vessel to help achieve the parabolic shape of the dish.

When deflated, the complete system fits in two airline-checkable cases that weigh less than 90 pounds each. The flexible antenna rolls up like a sleeping bag and fits in one case with the antenna bag, blower, hoses, and plates. The other case holds electronics, including a modem, spectrum analyzer, RF cables, computer, and power inverter.

"The fact that you can hand carry these communications terminals onto a roof and set them up rapidly is a main benefit of the technology," says Gierow, now the president of GATR Technologies. "If you tried to put a big rigid antenna on a roof, it would take several days."

The systems have been successfully deployed in missile ranges, severe weather, emergency response situations, and by soldiers at Fort Bragg, North Carolina. On a volunteer basis, GATR has used the communication systems to enable high-speed Internet and phone access during wild fires in southern California. After Hurricane Katrina, Gierow spent a week in Mississippi helping over 250 families and law enforcement officials with their communication needs. In January 2010, GATR provided Internet access to first responders and doctors following the earthquake in Haiti.

Used for contingency communication or backup communication when land lines are down or compromised, GATR systems have been deployed in the United States, Korea, Africa, Iraq, and Afghanistan. Popular with U.S. military and intelligence agencies, GATR recently signed a contract with the U.S. Navy, which will field the systems throughout the world.

The next step for GATR is to focus on two extremes—a small backpack version, and a large version that doubles the size of its current system. The small version aims to fit everything needed for a high-speed Internet connection into one backpack, including a fold-out solar array, communication gear, and antenna.

As for space-based inflatable antennas, the technology continues to evolve alongside ground-based systems, and will likely be fundamental to extended missions that require high data rates for space communication. ❖

Smart Sensors Assess Structural Health

Originating Technology/NASA Contribution

The materials used to make airplanes and space shuttles do not last forever. That is why NASA frequently inspects launch vehicles, fuel tanks, crew habitats, and other components for structural damage. The timely and accurate detection of cracks or other damage can prevent failure, prolong service life, and ensure safety and reliability.

The Composite Crew Module (CCM) is a full-scale test version of the Orion spacecraft. As seen here, the CCM has fiber optic and traditional strain gauges attached.

To perform quick, nondestructive evaluation and monitoring of aerospace vehicles and structures, NASA pursues the development of structural health monitoring (SHM) systems. SHM aims to build a system with a network of sensors placed in critical areas where structural integrity must be maintained, such as vehicle stage sections, separation interfaces, solid motors, and tanks. The sensors send information to a computer that is programmed to recognize patterns of electrical signals that represent damage, such as strains, breaks, or cracks. These systems can automatically collect and process data.

One approach to SHM is the Stanford Multi-Actuator Receiver Transduction (SMART) Layer, a patented technology from Stanford University. The thin material layer is embedded with a network of piezoelectric sensors and actuators that can be mounted on metal structures or embedded in composite structures. The piezoelectric sensors give off a small amount of electricity when they undergo mechanical pressure. Similar to a medical ultrasound, the sensors generate a wave that propagates through the structure and is picked up by other sensors. The aim is not only to detect structural damage, but to provide early warning before a failure takes place.

To enhance and commercialize the SMART Layer, Acellent Technologies Inc. was founded in 1999 in Sunnyvale, California. Soon after its founding, the company created an SHM system consisting of the SMART Layer, supporting diagnostic hardware, and data processing and analysis software.

Partnership

In 2001, Acellent started working with Marshall Space Flight Center through a **Small Business Innovation Research (SBIR)** award to develop a hybrid SMART Layer for aerospace vehicles and structures. The hybrid layer utilized piezoelectric actuators and fiber optic sensors. As a result of the SBIR, the company expanded the technology's capability to utilize a combination of sensors for various applications, such as monitoring strain and moisture.

A development known as the hybrid piezoelectric/fiber optic structural diagnostics system was intended to perform quick nondestructive evaluation and longer-term health monitoring of aerospace vehicles and structures, and could potentially monitor material processing, detect structural defects, detect corrosions, characterize load environments, and predict life. Piezoelectric actuators were embedded along with fiber optic detection sensors including Fabry-Perot fiber optic strain gauge sensors and fiber Bragg grating sensors.

Additional SBIRs with Marshall helped Acellent to improve and optimize its technology. The SMART Suitcase was developed and used for testing the SMART Layer and other SHM technologies. A NASA Space Act Agreement through Marshall investigated radio frequency attenuation transmission and detection using the SMART Layer approach. Marshall also provided knowledge advancement through testing opportunities with composite over-wrapped pressure vessels, which were being developed and studied for potential use in several space programs. Under a 2004 SBIR, Acellent tested and evaluated the performance of a SMART Tape, based on the SMART Layer, in harsh cryogenic conditions.

"We used the same base technology in all of the SBIRs, but for different applications requiring new innovations. This increased the reliability of the system and made it more robust," says Shawn Beard, chief technology officer at Acellent.

In 2009, Acellent's SMART Layer technology won the "Best Practical SHM Solution in Aerospace" award at the International Workshop on Structural Health Monitoring, sponsored by the Airbus Company. Twelve organizations participated in the competition, which was judged by representatives from industry, universities, and government agencies.

Beard attributes the technology's competitive advantage to the knowledge gained while working with

Marshall. "There are other technologies that are being applied for SHM, but they are behind in development of a complete system. There was a lot of testing and improving the technology over the last decade under the SBIR program to optimize the system," says Beard.

In 2009 and 2010, Marshall supported the implementation and testing of the Acellent pitch-catch piezoelectric sensor on the Composite Crew Module (CCM) for the Ares Program. The CCM is a full-scale test version of the Orion spacecraft.

Product Outcome

Acellent develops advanced active and passive diagnostic systems using built-in networks of actuators and sensors.

Customarily, SHM uses sensors and actuators arrayed at various locations on a structure. In contrast, the SMART Layer contains an entire sensor and actuator array, making it unnecessary to install each sensor and actuator individually. The SMART Layer is pre-networked and pre-positioned, making it very easy for installation. The layer can be mounted on an existing structure or integrated into a composite structure during fabrication. Different types of sensors, such as piezoelectric and fiber optic, can both be embedded in the sheet to form a hybrid network.

Capable of being integrated into new or existing structures, SMART Layer technology is used to automate inspection and maintenance including structural condition monitoring, load and strain monitoring, impact detection, damage detection, impact damage assessment, crack growth monitoring, debonding detection, process monitoring, materials cure monitoring, and quality control.

The main benefits of SMART Layer technology are its flexibility, light weight, ability to adapt to any structure, ease of installation, durability, and reliability under different environments. The size and shape of the technology varies, and Acellent manufactures SMART Layers in numerous sizes, shapes, and complexities—from single sensor flat strip to multisensor 3-D shells. In addition to the sensor network, Acellent provides the diagnostic hardware and data analysis software for different SHM applications from monitoring large composite structures to localized damage detection in composites and metals.

Acellent's SMART Layers can be customized with two or more sensors in a strip unit, like the eight-sensor strips shown here, to monitor structural health conditions such as strain and moisture.

Acellent has been funded to monitor civil infrastructure such as pipelines, buildings, and bridges.

According to Acellent, customers and partners include aerospace and automotive companies; construction, energy, and utility companies; and the defense, space, transportation, and energy industries. Private and government research labs have purchased the technology to gauge its performance in particular environmental conditions.

Acellent has had sales contracts with many major aerospace companies, including Boeing, Lockheed-Martin, Bombardier, and Airbus. Acellent is also customizing the technology for the U.S. Army to monitor fatigue cracks in rotorcraft structures. Recently, BMW used the SMART Layer technology on a concept vehicle to help verify the design and modeling, and to improve manufacturing techniques.

Over the last couple of years, Acellent has been funded to monitor civil infrastructure such as pipelines, buildings, and bridges, including a recent project with the National Institute of Standards and Technology. "We have been asked to develop a system—not just to monitor damage to a structure—but to monitor the overall framework or architecture for monitoring bridges across the Nation," says Beard.

In addition to winning sales for research and testing, the company has won additional NASA SBIRs to apply the technology to large scale structures like rocket motors. Acellent's long-term goal, however, is to have the technology become part of the design process. Instead of applying SHM after a new spacecraft is designed, Beard says the company envisions SHM being included in the original design of the structure. "If you design sensors into the structure in the beginning, you can optimize the structural design and reduce the overall weight of the vehicle. That is the direction we would like to see the technology go." ❖

SMART Layer® is a registered trademark of Acellent Technologies Inc.
SMART Suitcase™ is a trademark of Acellent Technologies Inc.

Hand-Held Devices Detect Explosives and Chemical Agents

Originating Technology/NASA Contribution

Smaller, with enhanced capabilities. Less expensive, while providing improved performance. Energy efficient, without sacrificing capabilities. Smaller, less expensive, and energy efficient—but still highly durable under some of the most extreme conditions known.

Contradictions like these are commonplace when designing sensor instruments for spacecraft. The Curiosity Mars Science Laboratory, for example, is set to launch in 2011 with an anticipated 10 instruments onboard that must endure the launch, an 8-month journey to Mars, landing, and the unfriendly environment of the Red Planet. Packing that much scientific punch into a single, reliable, one-shot package requires the most advanced technology available.

Given the high cost of development and launch; the need for consistent, high-level operation in harsh conditions and without the possibility of maintenance; and the limited real estate available on a spacecraft; NASA continually seeks ways both to develop new sensor technology and to advance existing devices to meet the demands of space exploration—in many cases through collaboration with private industry. Sometimes, the effort results in smaller, less expensive, more energy efficient, and highly durable products for use on Earth, as well.

Partnership

NASA's Astrobiology Science and Technology Instrument Development (ASTID) program encourages the development of innovative instruments equally capable of fulfilling astrobiology science requirements on space missions or related science objectives on Earth. Through ASTID and the **Small Business Innovation Research (SBIR)** program, Ion Applications Inc., of West Palm Beach, Florida, worked to meet NASA's need for a miniature version of a powerful sensor technology known as an ion mobility spectrometer (IMS).

Ion mobility spectrometry is a fast, highly sensitive method for separating and identifying gaseous molecules. In an IMS device, ionized molecules sampled from the air travel through a drift tube containing a buffer gas. The speed of travel is influenced by the ion's mass, size, shape, and charge. By measuring how quickly the ions migrate through the tube, IMS can identify a significant variety of molecules with part-per-billion sensitivity. The instrument displays lesser sensitivity toward other molecules, such as light hydrocarbons and noble gasses, under normal operating conditions.

"An IMS can basically detect almost anything, from heavyweight compounds all the way down to permanent gasses like hydrogen," says Alex Lowe, Ion Applications' vice president of sales and marketing. This capability makes IMS an obviously valuable detection instrument; in particular, it has been the technology of choice for explosives and chemical warfare agent detection since the 1960s and is in widespread use for airport security and military applications.

NASA had already developed a gas chromatograph ion mobility spectrometer (GC-IMS) called the mini-Cometary Ice and Dust Experiment (miniCIDEX). The device combines gas chromatography with ion mobility spectrometry to provide highly sensitive gas analysis for astrobiology missions. While the GC device had been successfully miniaturized, the IMS component needed to be smaller and capable of producing accurate readings with a reduced sample size. With SBIR support from Ames Research Center, Ion Applications developed a unique, miniature, Kovar (an alloy) and ceramic IMS cell; simplified electronics; and software for control and spectra acquisition.

The resulting Mini-Cell IMS proved to be a more sensitive, reliable, and rugged tool than existing IMS technology. For NASA, the improved IMS device could be used for future missions to planets, moons, and comets, or as a space-saving tool to monitor air quality on the International Space Station. On Earth, the applications are proving even more varied and beneficial.

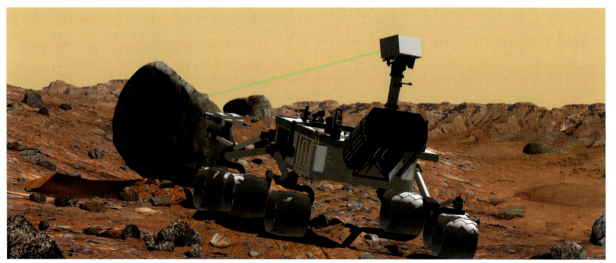

The Mars Science Laboratory, shown in the rendering above, will carry an anticipated 10 scientific instruments, the most for a Mars rover to date. These instruments must be rugged and powerful while still being able to fit within limited space on the rover platform.

Product Outcome

Ion Applications has commercialized the Mini-Cell IMS in the form of its EASYTEC-XP hand-held detector. Housed in a pistol grip casing with a point-and-shoot sampling mechanism, the EASYTEC-XP solves many of the existing engineering problems surrounding IMS devices, including portability.

"The traditional IMS has some inherent field problems that don't make it user friendly from a continuous operations standpoint," Lowe says. Unlike typical IMS devices, EASYTEC-XP does not require membranes to keep out moisture and other ambient air interference. Its IMS cell is 10 times smaller and more sensitive, and the detector requires no calibration and has annual maintenance costs that are about 8 times less expensive than the standard. It also requires no swipes or contact to make measurements, making it an ideal tool for field use.

"The traditional IMS products out there that don't use our platform aren't well suited for a number of applications, either in terms of size, price, or convenience," Lowe says. "We can basically plug into those applications right away."

Currently, the U.S. Army and Navy are using the EASYTEC-XP for detection of explosives and chemical warfare agents. Soldiers can simply carry the tool to a vehicle, for example, to inspect it for trace explosives or other dangerous chemicals. The device is also in use by international law enforcement, and the company has developed a narcotics detector capable of detecting heroin, cocaine, methamphetamine, and THC. EASYTEC-XP's versatility, durability, and portable format render it useful for transportation security, port container inspection, general air monitoring, and other military and security applications.

EASYTEC-XP has uses beyond public safety, as well. Ion Applications' technology is employed in China for high-purity process gas analysis for semiconductor manufacturing. The IMS readily detects damaging moisture in the special gasses used during the manufacturing process. Other industrial applications include cleanroom air monitoring and the detection of toxic gasses produced during industrial processes.

Ion Applications is continuing work on a wide range of potential applications for its NASA SBIR-derived technology, Lowe says.

"This is a platform that is allowing us to put together many commercial products that were never before attractive from an IMS standpoint." ❖

Kovar™ is a trademark of Carpenter Technology Corporation.
EASYTEC®-XP is a registered trademark of Ion Applications Inc.

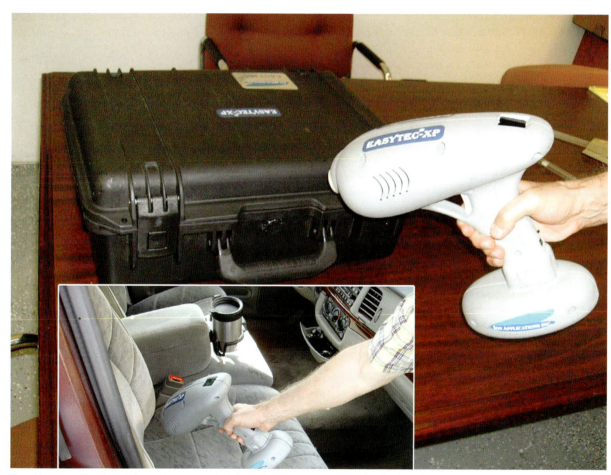

Requiring no calibration and no swipes or contact to make measurements, Ion Applications Inc.'s ion mobility spectrometer technology can inspect packages and vehicles for trace explosives and other dangerous chemicals. The device is in use by the U.S. military and international law enforcement.

Terahertz Tools Advance Imaging for Security, Industry

Originating Technology/NASA Contribution

On January 16, 2003, the Space Shuttle Columbia launched on mission STS-107. At T plus 82 seconds, with the orbiter rocketing upwards at 1,870 miles per hour, a briefcase-sized chunk of insulating foam broke off from the external fuel tank and struck Columbia's left wing. During reentry on February 1, hot gasses entered the wing through the damaged area of the orbiter's thermal protection system, causing devastating structural failure that led to the destruction of Columbia and the deaths of the seven crew members onboard.

After the Columbia disaster, NASA grounded the space shuttles for more than a year as it worked on new safety protocols to ensure that such a tragedy would not happen again. As part of the preparations for the Return to Flight mission, the Agency required a method for detecting potentially hazardous defects in the external tank's sprayed-on insulating foam prior to launch.

Partnership

NASA Langley Research Center scientists suspected that a new imaging technology called terahertz imaging had the potential to accurately find flaws in the foam on the external tank. Terahertz radiation—lying between microwaves and far infrared on the electromagnetic spectrum—offers imaging capabilities similar to X-rays, but unlike X-rays, terahertz radiation is non-ionizing and thus safe for frequent human use. Terahertz wavelengths can be used to see through many materials and reveal defects like cracks, voids, and density variations. They can be used to image or as an anomaly detector, or both at the same time.

Picometrix, of Ann Arbor, Michigan, was at the forefront of the emerging field of terahertz imaging. In 2000, Picometrix introduced the world's first commercial terahertz system, the T-Ray 2000. The T-Ray 2000 was based upon the company's patented fiber coupling system, but was a non-integrated, workbench-mounted

The space shuttle's external tank is coated in insulating foam that must be checked for defects. To accomplish this, NASA turned to the developing field of terahertz imaging.

Terahertz can be employed as a safer, more precise security measure than X-rays.

system, which rendered it fine for the research market but impractical for NASA's manufacturing quality control needs. Langley researchers asked the company via **Small Business Innovation Research (SBIR)** agreements to quickly redesign the terahertz systems to be more integrated and deployable into a manufacturing environment.

Based on the success of that new prototype system, the company was next asked to deliver a more compact, self-contained terahertz system, the T-Ray QA-1000, and NASA purchased five of the systems for inspecting the external fuel tanks as they were being manufactured by Lockheed Martin. The QA-1000's long, optical fiber umbilicals enabled the system's terahertz sensors to scan the tank from top to bottom. The systems were deployed at NASA's Michoud Assembly Facility and at Marshall Space Flight Center. Langley's original unit was later retrofitted with a similar higher speed delay stage that was also capable of imaging thicker foam.

"This was significant. In addition to the company's patented fiber coupling system that makes Picometrix systems unique, they can also inspect thicker material at substantially higher speed with our T-Ray systems versus others terahertz systems," says Irl Duling III, company director of terahertz business development.

Picometrix became a wholly owned subsidiary of Advanced Photonix Inc. (API), also of Ann Arbor, Michigan, in 2005. The company's terahertz systems—including its latest, highly compact and rugged T-Ray 4000 systems—were later adopted by Kennedy Space Center as a diagnostic tool for scanning the orbiter's thermal tiles for the remaining shuttle flights. The systems offered an effective way of not only inspecting the tiles for hidden damage, but also of precisely locating components underneath the tiles that were in need of attention—

without the costly removal and replacement of extra tiles which often happened before the use of the T-Ray 4000.

"With this technology, NASA could scan and see the precise location of wires and antennas and remove only the necessary tiles," says Picometrix engineer Greg Stuk. "In one example, it saved the Agency hundreds of thousands of dollars."

Product Outcome

The imaging capabilities of terahertz make it useful for a wide range of applications. It can be employed as a safer, more precise security measure than X-rays in airports and other buildings, revealing concealed weapons and the contents of packages. Since numerous materials have specific spectral signatures revealed by terahertz radiation, it provides spectroscopic and other unique identification information useful for chemical analysis, pharmaceuticals, and explosives detection. Not only can terahertz see through an opaque pill bottle, for example, it can also reveal the chemical makeup of the pills inside. It can also provide high-resolution imaging down to 200 microns. The industrial possibilities of terahertz range from determining the uniformity of coating thickness to detecting hidden defects to ensure product quality.

The company now offers the T-Ray 4000 Time-Domain Terahertz System commercially. Featuring its patented fiber-pigtailed transmitter and receiver modules, the T-Ray 4000 is designed for both the research laboratory and the industrial setting. The T-Ray 4000 takes the next step beyond the NASA-inspired T-Ray QA-1000 system. While the QA-1000 is about the size of a small refrigerator, the T-Ray 4000 is an easily portable, rugged, briefcase-size system weighing only about 50 pounds. As a time-domain terahertz system, the T-Ray 4000 generates high-speed picosecond (one-trillionth of a second) duration terahertz pulses for scanned spectroscopy or imaging. These qualities, along with the patented fiber-coupled sensor heads that can scan objects of almost any size, make the T-Ray 4000 an easy-to-use tool for terahertz applications beyond the laboratory—though it is useful there as well.

"As far as having a product that you can deploy onto a manufacturing floor, this is the first of its kind," says Duling. He credits the company's NASA work with helping drive this industry-leading advancement.

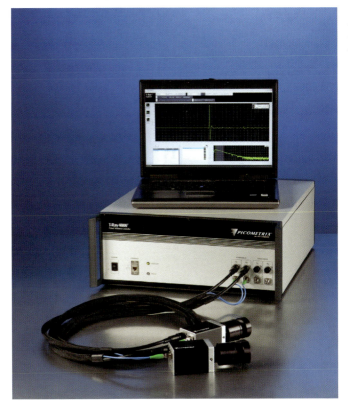

The T-Ray 4000 Time-Domain Terahertz System, capable of generating high-speed terahertz pulses and equipped with fiber-coupled sensor heads, is a versatile and user-friendly instrument for use in the laboratory and in real-world applications.

"The rest of the industry is trying to figure out how to generate terahertz, how to detect it, how to build a complete system that can be fielded," he says. "In part through NASA's motivation, we've been able to complete that full-system integration and turn it into something we can take out into the field and use as a tool."

"Our systems' features now allow terahertz to access the most obscure places," says Steve Williamson, the company's chief technology officer. "It's a powerful benefit to our customers."

API's terahertz systems can be used for thickness measurements of roofing material, paper and paper coatings, and coatings on films. They also can be employed for pharmaceutical applications like aseptic packaging and tablet production. Art conservationists from prestigious institutions like the Uffizi Gallery and the Louvre have used API systems to date paintings, look for pigment concentrations, and reveal frescos on walls that have been painted over. The technology has been even applied to examine the structure of pagodas in Japan, providing guidance for renovations. These are only a few examples of the benefits of this still developing field, says API CEO Richard Kurtz.

"Terahertz has huge market potential. We estimate there are over $200 million in opportunities for our terahertz systems over the next 7 years," he says. To help API stay at the forefront of the terahertz industry, the company is continuing work with NASA through SBIR contracts with Glenn Research Center. The goal of this partnership is a computed axial tomography time-domain terahertz system capable of creating three-dimensional images.

"There has been great collaboration between API and NASA," says Williamson. "NASA has helped us push the envelope." ❖

T-Ray®, T-Ray 2000®, and T-Ray 4000® are registered trademarks of Advanced Photonix Inc.

Consumer Goods

NASA's aerospace research often comes back to Earth in surprising but very practical ways, improving the quality of our everyday lives. The technologies featured in this section:

- Target Plant Growth
- Insulate Against Extreme Temperatures
- Enhance Camera Technologies
- Allow for Quick Repairs
- Transform Hairstyling Tools
- Recharge Auto Air Conditioning

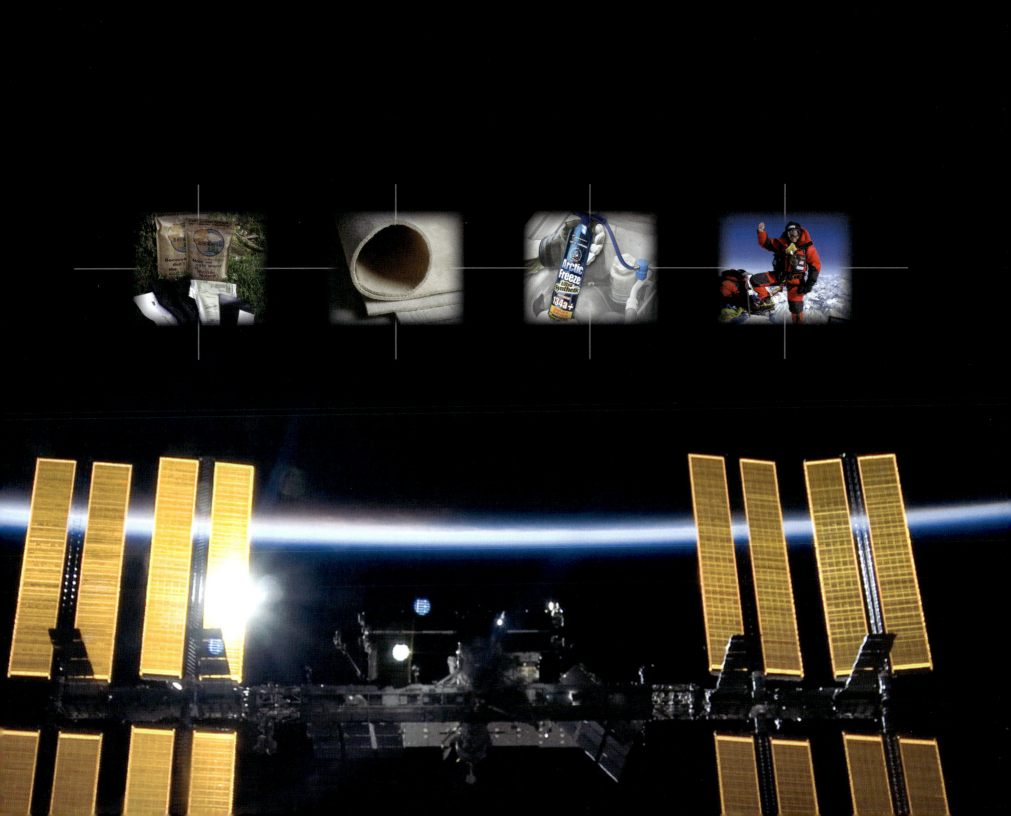

LED Systems Target Plant Growth

Originating Technology/NASA Contribution

Living in space long term will require a sustainable environment. Plants provide fresh food, clean air, and clean water that will assist this effort, but plants need light to grow, and light requires energy. Here on Earth, most plants get this light from the obvious abundant source, the Sun. The Sun's solar radiation is ideal for growing plants here on Earth, but it presents some problems for plant growth in space. For starters, the lengths of the days are different depending upon the location of the garden. For growing plants on spacecraft, this problem is compounded, as the vehicle position is constantly changing and is usually not positioned for optimal plant growth. Thus, NASA has been developing methods for growing crops in space using artificial light sources.

Lighting plants with electric lamps overcomes several difficulties, but presents an additional problem: It can require a great deal of valuable energy and produce unwanted heating of the plants. The solution appears to be LED lighting, which allows for precision and control, uses less power than traditional lamps, and radiates minimal heat onto the plants. In addition, LED lighting typically lasts much longer than traditional bulbs, is smaller and lighter weight, and does not present the same potential risks of glass breakage as traditional bulbs.

Partnership

To help develop technologies for growing edible biomass (food crops) in space, NASA partnered with a small business in Wisconsin. Orbital Technologies Corporation (ORBITEC), located in Madison, is one of the state's leading developers of new and cutting-edge aerospace technologies. The company has been awarded over $125 million in government contract funds over the course of more than 180 contracts, most of which were through commercial contracts that began in the **Small Business Innovation Research (SBIR)** program. Over the course of its extensive government work, ORBITEC was awarded Wisconsin's "Professional Service Business of the Year" in 1995 and received the "Tibbets Award" both in 1996 and 1999 from the Small Business Administration. The "Tibbets Award" acknowledges small businesses that have performed exceptionally well within the SBIR program. One of the highlights of ORBITEC's space-related work includes the 2002 launch of its Biomass Production System aboard STS-110 for a 73-day stay aboard the International Space Station.

One of the recent projects ORBITEC has been working on with Kennedy Space Center is the development of the High Efficiency Lighting with Integrated Adaptive Control (HELIAC) system, which uses targeted solid-state LEDs to efficiently grow plants. One configuration of the HELIAC system consists of a series of LED light panels called light engines. About 4 cm square and arranged in rows called "lightsicles," these light engines are precision-controlled and allow for maximum efficiency in plant growth. Research for space applications is continuing through a partnership between ORBITEC and Purdue University.

While NASA is keen on this promising technology for future experiments in space, hardware and software protocols developed through the HELIAC program have the potential to save energy in commercial agriculture and in aquarium lighting while providing a host of additional benefits.

The NASA-derived light distribution systems are low power, relatively cool, uniformly irradiate all leaves with wavelengths most efficiently absorbed by photosynthetic tissue, and automatically adjust emissions to target new tissues as plants grow in height or spread, without wasting photons by lighting empty space.

Product Outcome

LED lighting systems are robust, easy to maintain, require less energy, produce little radiant heat, and reduce the dangers associated with pressurized bulbs, broken glass, mercury and high surface temperatures. What makes them truly ideal for plant growth, though, is their variable light output control. ORBITEC's precision HELIAC control system allows lamp configuration to be adapted to a specific plant species during a specific growth stage, allowing maximum efficiency in light absorption by all available photosynthetic tissues. It can sense the presence of plant tissue and only power the adjacent elements, thus providing efficient, targeted lighting. Picture, for example, a newly sprouted plant. Traditional lighting for this plant would be broad and scattered mostly over the growing medium. The HELIAC system is able to sense exactly where the plant's leaves and shoots are and spotlight those areas. This significantly reduces energy usage.

So far, ORBITEC has implemented a number of the technologies developed under the HELIAC project—several advanced control algorithms, sensors, drive circuits, thermal systems—into its commercial products. For example, horizontal light bars that allow real sunlight between the bars when available have been sold to research universities and controlled environment system manufacturers. The supplemental lighting allows growers to take full advantage of natural sunlight while also providing targeted lighting if sunlight is not available. This allows, for example, the plants to continue getting necessary light even on cloudy or rainy days.

The University of California, Davis purchased 216 of ORBITEC's greenhouse bars for photobiology research, while many other customers are purchasing these units on a smaller scale for research or evaluation of the new technology for full-scale implementation. The units are still manufactured in-house by ORBITEC, but this may change as demand increases. Currently, though, the technology is constantly evolving.

Another application ORBITEC has found for its controllable plant growth technology is in aquarium lighting. In aquariums, water temperature is a major issue—it has to be just right for certain fish or corals to thrive. With a traditional light fixture, an aquarium also needs to be equipped with a chiller to keep the water from heating. With LEDs, which produce minimal radiant heat towards the water, this is not a problem. They also last much longer than traditional bulbs.

The primary advantage of the ORBITEC system over other LED-based aquarium lighting systems, however, is the programmability. Using drive circuitry, software, and thermal protocols developed for the NASA work, these lights offer unprecedented adaptability and precise controls. They can be set to provide a specific spectrum depending upon the contents of the aquarium. Subtle spectral adjustments can be made for coral or specific fish. The lights can even be set, for example, to dim periodically to create the illusion of a cloud passing over head, the changes of the tide, or to simulate the lunar cycle. These sophisticated settings allow aquarium life to grow and thrive as if it were in a natural environment.

Currently, ORBITEC has licensed this technology to two aquarium lighting manufacturers, one of which, C2 Technologies Inc., is manufacturing the devices and marketing under the name AquaIllumination. ❖

AquaIllumination™ is a trademark of C2 Technologies Inc.

These LED systems substantially lower energy costs of controlled-environment production and will improve profitability.

Aerogels Insulate Against Extreme Temperatures

Originating Technology/NASA Contribution

"When you hold a piece of silica aerogel, it feels otherworldly. If you drop it on a table top, it has an acoustic ring to it. It sounds like a crystal glass hitting the table," describes George Gould, the director of research and development at Aspen Aerogels Inc.

Similar in chemical structure to glass, aerogels have gas or air in their pores instead of liquid. Developed in the United States nearly 80 years ago by a man named Samuel Stephens Kistler, an aerogel is an open-celled

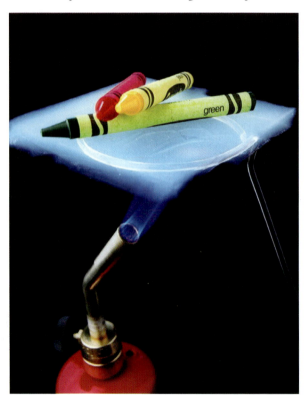

Crayons placed on top of a piece of silica aerogel will not melt from the heat of a flame. Certain types of aerogel provide 39 times more insulation than fiberglass.

Aerogel is made from a wet gel that is dried. The substance has been described as feeling like volcanic glass pumice; a very fine, dry sponge; and extremely lightweight Styrofoam.

material that is typically comprised of more than 95 percent air. With individual pores less than 1/10,000th the diameter of a human hair, or just a few nanometers, the nanoporous nature of aerogel is what gives it the lowest thermal conductivity of any known solid.

The remarkable characteristics of silica aerogel—low density, light weight, and unmatched insulating capability—attracted NASA for cryogenic insulation for space shuttle and space exploration mission applications. For example, when a shuttle is fueled, it requires more than half a million gallons of cryogenic liquid oxygen and liquid hydrogen. To remain a liquid, hydrogen must stay at a cold -253 °C and liquid oxygen must remain at -183 °C. The systems necessary to deliver, store, and transfer these cryogenic liquids call for high-performance insulation technology at all steps along the way and into space.

Aspen Systems Inc. worked with NASA to manufacture a more durable form of aerogel. The flexible material is made by filling the spaces of a fiber web with silica aerogel.

In 1992, NASA started to pursue the development of a practical form of aerogel. Up until that point, aerogel had always been too fragile to handle in its monolithic (or solid) form, and too time-consuming and expensive to manufacture. The concept for a flexible aerogel material was introduced by James Fesmire, the senior principal investigator of the Cryogenics Test Laboratory at Kennedy Space Center. Fesmire, at that time a mechanical engineer responsible for cryogenic fueling systems design, envisioned an aerogel composite material that would be practical to use, but would still exploit the phenomenal heat-flow-stopping capability provided by the nanoporous aerogel.

Partnership

Kennedy Space Center awarded Aspen Systems Inc., a research and development firm in Marlborough, Massachusetts, a **Small Business Innovation Research (SBIR)** contract to create a flexible, durable, easy-to-use form of aerogel. The world's first aerogel composite blankets were produced in 1993 as cookie-sized laboratory specimens. Initial testing under cryogenic conditions showed the material to have exceptionally good insulating performance in ambient pressure environments. At that time, standard laboratory test machines were inadequate to fully characterize the material's very low heat transfer characteristics under cryogenic conditions. A second phase of the SBIR program, a collaborative effort with Kennedy, was awarded in 1994. As part of that collaboration, a cryostat insulation test apparatus was devised for measuring the true thermal performance of the aerogel blankets. This apparatus, Cryostat-1, was able to fully test the material and later became the cornerstone capability for the laboratory at Kennedy.

By 1999, these contracts led to further partnerships, and Aspen Systems developed a manufacturing process with NASA that cut production time and costs, as well as produced a new form of aerogel, a flexible aerogel blanket. To make the new material more useful, the spaces within a web of fiber reinforcement were completely filled with silica aerogel. "It's a little like an epoxy resin in the polymer composites world. By itself, epoxy resins can make great glue. But if you combine it with fiber, you can make airplanes and helicopters out of it," says Gould.

To develop and market the revolutionary product, Aspen Systems started Aspen Aerogels Inc. in Northborough, Massachusetts. Since 2001, Aspen Aerogels has been using the same manufacturing process

Aspen Aerogels Inc. produces nearly 20 million square feet of aerogel material per year and sells it for government, industry, and consumer applications.

developed in part under the NASA SBIR to provide aerogel to the commercial world. In 2003, Aspen Aerogels received the "R&D 100" award from *R&D Magazine*. By 2009, the company had become the leading provider of aerogel in the United States and currently produces nearly 20 million square feet of the material per year.

Product Outcome

While NASA uses Aspen Aerogels' product for cryogenic applications such as launch vehicles, space shuttle applications, life support equipment, and rocket engine test stands, there is an array of commercial industrial applications including pipe insulation, building and construction, appliances and refrigeration equipment, trucks

A company called Polar Wrap LLC encapsulates the NASA-derived aerogel and uses it in a product called Toasty Feet (shown above and on the right). These insoles protect people's feet from both heat and cold.

Mountaineer Ann Parmenter summitted Mt. Everest on May 25, 2006. She said her feet stayed comfortable and warm while wearing just one pair of socks—plus Toasty Feet insoles—inside her climbing boots.

and automobiles, as well as consumer applications, such as personal apparel. Most recently, the NASA-derived aerogel has been applied to protect and insulate people's hands and feet.

Polar Wrap LLC, is a Memphis, Tennessee, company that buys the material from Aspen Aerogels and then applies its own patented process to encapsulate the aerogel and use it in insoles called Toasty Feet. Designed to fit in the bottom of a boot or shoe, Toasty Feet resists heat loss and heat gain. According to the company, sales totaled over a million and a half pairs in 2009. Their line of insoles includes mens, womens, youth, extra cushion, and arch support.

The inventor of the process to encapsulate the aerogel for Polar Wrap was originally looking for insulation for the refrigeration system on his sailboat. When he saw the capabilities of aerogel, he thought the material held promise for the company. The inventor then devised an application for clothing, which resulted in the process now used to make Toasty Feet.

According to Polar Wrap, two people walked the length of the Great Wall of China (a 4,500-kilometer walk that took 6 months) wearing Toasty Feet. A mountaineer climbed Mount Everest using Toasty Feet instead of liner socks and said her feet stayed warm. In addition, an endurance runner who ran a marathon from Death Valley to Mt. Whitney, California, said her feet stayed heat-free while wearing Toasty Feet.

Another company looking for ways to warm feet—and hands—also decided to use Aspen Aerogels' product. Originals By Weber, of Toms River, New Jersey, is an Internet-based business. The owner, Terrance L. Weber, wanted a way to help people with Raynaud's disease, a condition that causes the fingers and toes to feel numb and cool in response to cold temperatures or stress. The smaller arteries that supply blood to the skin become narrow, limiting the blood circulation to affected areas.

To keep the blood warm, Weber decided to try applying insulation to the wrists and ankles. After experimenting with several materials, including a fiberglass

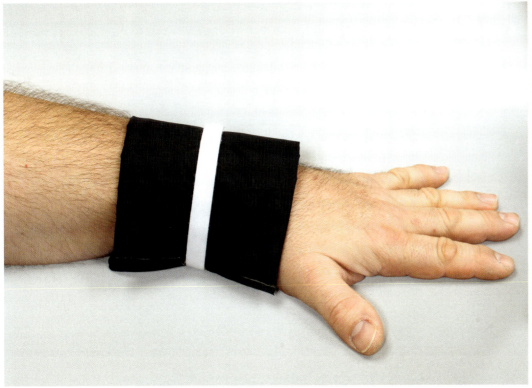

The Wrist and Ankle Wraps (to the left and above) were made by Originals By Weber to help people with Raynaud's disease fight painfully cold hands, fingers, feet, and toes. According to the company, ultra-thin aerogel insulation assists in controlling and maintaining blood temperature, and also increases blood flow to the hands and feet.

product, he says, "I chose aerogel because it is thin and lightweight, and almost to the point where you don't even know it is there."

Encased in nylon, the Wrist and Ankle Wraps are secured with a strap to maintain the normal temperature of the blood as it flows from wrists to hands and fingers, and from ankles to feet and toes. In the course of 6 months, the company has sold about 75 pairs of the product.

In addition to insoles, and wrist and ankle wraps, the NASA-derived product has also made its way into boots. Salomon, a French company that sells sporting products, incorporates aerogel into its Toundra winter boots for men and women. Another French company, Heckel, incorporates aerogel insulation from Aspen Aerogels in its MACPOLAR boots. The company ensures comfort in temperatures as low as -50 °C, and promotes the boots for refrigerated warehouses, oil and gas exploration, snow and ski slope services, mines, transport services, and other harsh winter conditions.

Many new applications are on the horizon for space applications as well. The aerogel blanket material is enabling new ways of designing high-performance systems of all kinds for extreme environments. The atmospheres of Earth, the Moon, and Mars all present unique challenges for controlling and saving energy. With applications across various industries, Gould traces much of aerogel's commercial success to working with NASA early in the development cycle. "If you can meet NASA's high expectations for performance and safety requirements, and subsequently make a product that has commercial potential, you are on a great path to delivering goods that are the best in class." ❖

Toasty Feet™ is a trademark of Polar Wrap LLC.
Styrofoam™ is a trademark of The Dow Chemical Company.

Image Sensors Enhance Camera Technologies

Originating Technology/NASA Contribution

Buzz Aldrin standing on the stark surface of the Moon. The towering gas pillars of the Eagle Nebula. The rocky, rust-colored expanses of Mars. Among NASA's successes in space exploration have been the indelible images the Agency's efforts have returned to Earth. From the Hubble Space Telescope to the Hasselblad cameras in the hands of Apollo astronauts, many of NASA's missions involve technologies that deliver unprecedented views of our universe, providing fuel for scientific inquiry and the imagination.

Less known than Hubble's galactic vistas or the Mars rovers' panoramic landscapes is the impact NASA has had on the era of digital photography on Earth. While the first digital camera was built by Eastman Kodak in 1975, the first to actually develop the concept of the digital camera was Jet Propulsion Laboratory (JPL) engineer Eugene Lally, who in the 1960s described the use of mosaic photosensors to digitize light signals and produce still images. During the following decades, NASA continued the work of developing small, light, and robust image sensors practical for use in the extreme environment of space.

In the 1990s, a JPL team led by Eric Fossum researched ways of improving complementary metal-oxide semiconductor (CMOS) image sensors in order to significantly miniaturize cameras on interplanetary spacecraft yet maintain scientific image quality. An image sensor contains an array of photodetectors called pixels that collect single particles of light, or photons. (The word "pixel"—short for picture element—was first published in 1965 by JPL engineer Frederic Billingsley.) The photons entering the pixel are converted to electrons, forming an electrical signal a processor then assembles into a picture. CMOS sensors represented a number of appealing qualities for NASA compared to the charge coupled device (CCD), the prevalent image sensor at the time. Crafted by the same process used to build microprocessors and other semiconductor devices, the CMOS image sensors can be

Images like these of (clockwise from top left) the Eagle Nebula, Apollo 11 astronaut Buzz Aldrin on the Moon, the Martian landscape, and an astronaut self-portrait taken during a spacewalk bring the wonders of space exploration down to Earth.

manufactured more easily than CCDs and at lower cost. The CMOS sensor components are integrated onto a single chip, unlike CCDs, which have off-chip components. This integrated setup consumes as much as 100 times less power than CCDs, allows for smaller camera systems, and can be designed with radiation-hard pixel architectures for space applications.

At JPL, Fossum invented the CMOS active-pixel sensor (CMOS-APS), which integrates active amplifiers inside each pixel that boost the electrical output generated by the collected photons. The CMOS-APS featured improved image quality over passive-pixel sensors (without amplifiers) and included a number of on-chip functions, providing for complete miniature imaging systems that operate quickly with low power demands. JPL validated the technology through a series of prototypes.

Partnership

Fossum realized the CMOS-APS technology would be useful not only for imaging in space but on Earth as well. In 1995, he, his colleague and then-wife Sabrina Kemeny, and three other JPL engineers founded Photobit, based in Pasadena, California. Photobit exclusively licensed the CMOS-APS technology from JPL, becoming the first company to commercialize CMOS image sensors.

"We saw an expanding number of applications for these miniaturized cameras," says Roger Panicacci, one of Photobit's founders. The company quickly positioned itself on the cutting edge of the field of CMOS imaging, and by June 2000, it had shipped 1 million sensors for use in popular Web cameras, machine vision solutions, dental radiography, pill cameras, motion-capture, and automotive applications. The company was featured in *Spinoff* 1999 and founders Fossum, Panicacci, Kemeny, and Robert Nixon were inducted into the Space Foundation's Space Technology Hall of Fame that same year.

In 2001, the company was acquired by semiconductor memory producer Micron Technology, of Boise, Idaho, and became a division of Micron Imaging Group. With the exploding popularity of the camera phone in the mid-2000s, the CMOS-APS proved ideal for crafting cameras that fit into slim cell phones and produce good photos without draining batteries. Riding the wave of camera phone demand, in 2006 the group became the world's leading supplier of CMOS image sensors. In 2008, Micron Imaging Group was spunoff from Micron to form Aptina Imaging Corporation, based in San Jose, California. That same year, it shipped its 1 billionth sensor.

Product Outcome

Aptina has continued to improve on the original, NASA-developed CMOS-APS. The company has invented increasingly small pixel architectures, as well as a process for optimizing the amount of light that hits a pixel, boosting sensitivity and image quality while allowing the company's customers to design more compact camera systems.

"Our technology is taking advantage of semiconductor innovation," says Panicacci, now Aptina's vice president of product development. "As transistors shrink, we can build smaller pixels, meaning that, in a given area of silicon, we can provide higher and higher resolution for products like camera phones."

Aptina's line of sensors allows for advanced camera features like electronic pan, tilt, and zoom, as well as

Aptina Imaging Corporation's 10-megapixel complementary metal-oxide semiconductor sensor is the first such device for point-and-shoot cameras.

applications requiring motion detection, target tracking, and image compression. The sensors are also incorporated into the company's line of system-on-a-chip (SOC) devices—synergistic packages that enhance imaging, are easier and cheaper to integrate into products, and provide benefits like anti-shake compensation that corrects blurring from subject motion or an unsteady camera.

Aptina has grown from its NASA roots into a leader of the CMOS image sensor industry. Its sensors are currently integrated into one of every three cell phone cameras and are part of every major brand personal computer camera worldwide, as well as many embedded cameras for notebook computers. The company is also advancing CMOS sensors for digital still and video cameras—products that have traditionally featured CCD sensors, and has produced the first 10-megapixel CMOS image sensor for point-and-shoot cameras, a device that incorporates the company's High Speed Serial Pixel Interface (HiSPi) capabilities, enabling a camera to create high-definition (HD) imagery.

The NASA-derived CMOS-APS can also be found in other, less obvious applications. Aptina produces tiny sensors for use in endoscopes for minimally invasive medical diagnostic procedures. The sensors do not generate potentially painful heat during examinations and are cheap enough to allow for disposable scope tubes, eliminating potential complications from improperly sterilized scopes. The company also worked with a medical imaging partner to develop the PillCam, an ingestible camera for imaging a patient's gastrointestinal tract.

The automotive and surveillance industries represent other major markets for Aptina's sensors. Major international auto brands like Daihatsu and Volvo employ Aptina designs for applications like backup cameras that help with parking and ensure safe reverse motion. Aptina estimates its customers will be building up to 25 million backup camera systems annually by 2011, potentially reducing backover accidents by about 20,000 per year. The company also has partnerships within the field of network surveillance, designing imaging technology that can spot cheating in casinos or intruders in unauthorized areas.

Last year, Aptina became a stand-alone company, and while Photobit once saw 1 million sensors shipped as a major milestone, says Panicacci, Aptina often ships over 1 million sensors a day. As demand rises for high-end capabilities like HD imaging and the market for camera products booms, Aptina's NASA-developed technology should play an even greater role in products benefiting the public every day. Research by the International Data Corporation, an independent market research company, predicts annual sales of more than 1 billion camera phones beginning in 2010, all featuring one or more CMOS-APS cameras. ❖

HiSPi™ is a trademark of Aptina Imaging Corporation.
PillCam® is a registered trademark of Given Imaging Ltd.

Lightweight Material Patches Allow for Quick Repairs

Originating Technology/NASA Contribution

Here on Earth, if your sink springs a leak, you can call in a plumber, or if you're handy, you can head out to the local hardware store, buy a few replacement parts, and fix the problem yourself. If the leak isn't particularly bad, you can even place a bucket under the sink to catch the dripping water and put the chore off until the weekend. These options aren't exactly available to astronauts working on the International Space Station. They can't call in a specialist to make repairs when problems occur, and they can't run out to the hardware store for the exact parts needed for a repair. Plus, there isn't much free time in an astronaut's onboard schedule. Repairs need to be made as soon and as efficiently as possible. Toward that end, NASA funded the design of simple and reusable patch repair systems for servicing, maintaining, and repairing structural components in space without the need for heavy machinery or an expense of time.

Partnership

Cornerstone Research Group Inc. (CRG), of Dayton, Ohio, works in a variety of fields to produce high-tech solutions and provide technology development services. CRG specializes in transitioning new ideas from the laboratory to the market, which made it a good fit for working with NASA. It has been the recipient of 16 **Small Business Innovation Research (SBIR)** contracts with NASA, with a variety of different focuses, including projects like creating inflatable structures for radio frequency antennas and, most recently, healable polymer matrix composites for future space vehicles. One of its earlier SBIR contracts, with Kennedy Space Center, led to the development of a new type of structural patch for a variety of consumer uses. While this particular project only ran through a Phase I contract with NASA, according to CRG's Brenda Hood, "So much happened during that initial Phase I research that we knew we had a product with a lot of commercial value."

CRG Industries LLC, of Dayton, Ohio—a spinoff company of Cornerstone designed to manufacture and distribute the state-of-the-art materials developed by its parent company and specializing in moving the advanced research into the consumer markets—has commercialized the NASA-derived material under two trademarked names: Rubbn'Repair, for automotive uses; and Rec'Repair for the outdoors and adventure market.

Rec'Repair is a tough, formable patch for easily repairing holes and damage to aluminum, steel, other metals, fiberglass, glass, painted surfaces, plastic, some wood, stiff vinyl, copolymers, and composites.

Product Outcome

The Rubbn'Repair patch is tailored for automotive use, providing rigid, strong repairs for holes and damage to body panels, fenders, and bumpers, with the capability to even replace missing structural or body material. Once the adhesive patch is heated to approximately 194 °F, it becomes flexible and moldable and can be applied to the damaged area. When the material cools—in seconds—it becomes a rigid structural patch.

In auto racing applications, where the cars are exposed to constant stresses and repairs need to be conducted as quickly as possible, the patch makes for an ideal temporary fix. Since the adhesive can be heated multiple times, pit crews can store the nontoxic patches ready for use, continually applying moderate heat to keep the product flexible until use. Once the patch is applied, the car can be back on the track in a matter of seconds, with the patch able to withstand the vibrations and stresses of high speeds. The Rubbn'Repair patch is a structural composite, and unlike typical racing tape, will not degrade and delaminate at high speeds, providing a quick option for restoring aerodynamic shaping and preventing further damage.

It is also available for use in standard auto repair, where once it has been applied, the material can be machined and painted, providing a structural alternative to the thin patches usually used in bodywork. It can also be used behind a bumper, providing rigid adhesion and structure while a hole is repaired with conventional fillers. Once cured, the patch will not degrade, crack or crumble; it is waterproof, UV-resistant, vibration resistant, and can withstand typical weather-related temperature fluctuations. Rubbn'Repair has been particularly embraced by the trucking industry, where quick repairs on the road allow goods to be delivered on time, and pulling over on the side of the road to patch a leak in a refrigerated truck's insulation or hold a broken piece of fender on long

The patch becomes rigid after it is shaped and then cooled, creating a durable, water-resistant, structural repair that does not crumble or peel.

Heated to about 194 °F, the material becomes flexible and moldable to any repair surface. As it cools, it becomes rigid again and retains the new shape. At operating temperatures, the material is rigid and strong.

enough to get the product delivered is highly preferable to stopping for professional repair.

Based on the same technology, CRG created another application for the composite material: a portable patch for a variety of outdoors and emergency applications. Rec'Repair is a fully cured repair patch, consisting of high-tensile fabric, the patented shape-changing resin, and an industrial-strength adhesive surface on one side. When heated, the patch becomes pliable and can be shaped to fit a wide assortment of angles. The patch becomes rigid when it cools, providing a water- and weather-resistant solution that is stronger than duct tape, and takes fewer than 10 minutes to apply.

Rec'Repair is designed to meet the needs of campers, boaters, vacationers, and adventurers when they need an emergency, temporary structural patch, whether it is for a cooler, fishing rod, or even the bottom of a canoe. It is lightweight and easily packed and carried, and allows for broken or damaged gear to be repaired on the fly, with little delay.

Low levels of heat, such as those provided by a hot air gun or even a short period in the microwave, are enough to make the patch flexible. For field use, though, the patch comes with a water-activated heating pouch similar to those used in military "Meal, Ready to Eat" (MRE) packaging that, in 10 minutes, warms the patch enough to become flexible for application. Once the patch is flexible, the user removes the adhesive backing, and then applies the patch to the desired area, pressing firmly with just hand pressure. After the patch cools, the damage is repaired.

Applications for this space-derived, portable, moldable patch are nearly limitless. According to Hood, "People know that NASA conducts a lot of high-tech, cutting-edge experimental work, and that oftentimes products develop out of this work. This is one of those products." ❖

Rec'Repair® and Rubbn'Repair® are registered trademarks of Cornerstone Research Group Inc.

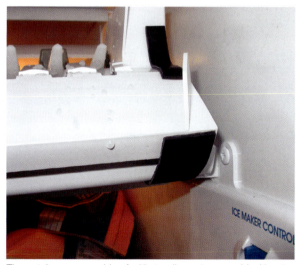

The patches are capable of withstanding extreme cold without breaking down.

Nanomaterials Transform Hairstyling Tools

Originating Technology/NASA Contribution

Dr. Dennis Morrison, a former scientist at Johnson Space Center, spent part of his 34-year career with NASA performing research on nanomaterials—materials 10,000 times smaller than a human hair. Specifically, Morrison's research on nanoceramic materials started with the development of microcapsules, or tiny balloons the size of blood cells, designed to deliver cancer-fighting drugs by injection into solid tumors deep within the body.

Originally, these liquid-filled microballoons were made in low Earth orbit where the absence of gravity aided in the formation of the outer membrane. Eventually, these space-based experiments resulted in the development of a device that could make the drug-filled microcapsules on Earth.

In order to release all of the contents of a microcapsule when and where a physician wanted them, Morrison developed special ceramic nanoparticles containing a unique mixture of metal oxides. When these metallic-ceramic composites were incorporated into the surface of the microcapsule and subsequently heated by a magnetic field such as a Magnetic Resonance Imaging (MRI) diagnostic device, the composites produced negative or positive ions and heated to a predictable temperature. This caused the tiny ceramic-magnetic particles to melt holes in the microcapsule, and thereby release the drug contents on command. Eventual tests on human tumors grown in mice showed three small injections of the microcapsules within a 2-week period inhibited more than 50 percent of the tumor's growth.

Partnership

To exchange and share research on nanomaterials, Morrison attended nanotechnology conferences sponsored in part by NASA. At a 2001 conference in Galveston, Texas, Morrison met Farouk Shami, the founder and chairman of Farouk Systems Inc., a manufacturer of professional hair care and styling products in Houston. Shami asked Morrison about the latest ceramic applications in nanotechnology because he wanted to enhance the ceramic materials used in his company's professional hairstyling tools. Specifically, Shami had developed slick ceramic coatings that emitted negative ions when heated.

After learning more from Morrison about ceramics containing special metal composites, Shami developed a unique ceramic composite and incorporated the material into his CHI (Cationic Hydration Interlink) hairstyling iron. When heated at low temperatures—roughly 180 to 200 °C—the metal components in the CHI ceramic released ions that proved beneficial for hairstyling. According to the company, the ions help to smooth and soften the hair, thus making it easier to manage and style.

"At Johnson, we were developing ceramic metallic components for triggering the release of drugs from microcapsules. I never had any idea that it might be beneficial to someone in the hair industry making a hair iron with ceramic plates," says Morrison.

Then in 2004, another area of NASA research inspired Shami to develop his products further. Because typical cleaning agents like chlorine and alcohol release fumes in the contained environment of a spacecraft, scientists and engineers at Johnson had looked at alternative methods to keep surfaces clean and disinfected. One of these methods involved the use of nanosilver particles. Because nanosilver acts as a passive sterilizing component, Shami found that by incorporating it into CHI hair tools and products, it essentially created tools with self-disinfecting surfaces. Airborne microbes like fungi and bacteria died after they settled on the surface. This feature was especially attractive to salons and spas, where emphasis is placed on improved air quality, reduced fumes, and reduced contamination.

Product Outcome

Today, Farouk Systems manufactures approximately 1,000 products including hairstyling tools, skin and hair conditioners, hair color for professional salons, and exports to 106 countries. In addition to using nanomaterials in the CHI hair iron, the company also uses the materials in an extensive line of brushes, curling irons, CHI nail lacquers, and hair dryers. The nanoceramic metallic composite in the outlet of the hair dryers releases the same beneficial ions and infrared wavelengths as the CHI hairstyling irons.

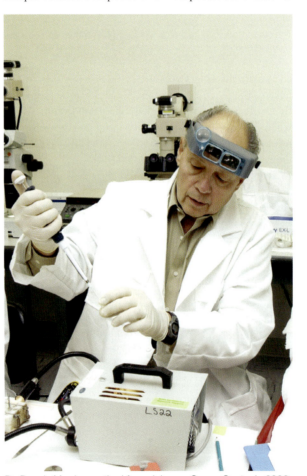

Dr. Dennis Morrison retired from Johnson Space Center in 2006.

To extend the potential of the combination of ions and infrared wavelengths created by CHI styling tools, Shami created a second level of products based on the NASA-inspired technology—liquid formulations specially designed to work with the irons and blow dryers. These include the CHI Ionic Color Protector System, the CHI Curl Preserve System, and the CHI Organics Olive Nutrient Therapy line, among others. Shami also invented special hair conditioners containing natural silk to use with the heated ceramic plates to prevent hair from burning or drying out. According to Farouk Systems, the infrared thermal waves plus silk infusion not only help to seal natural oils in the hair, but improves moisture retention during the styling process.

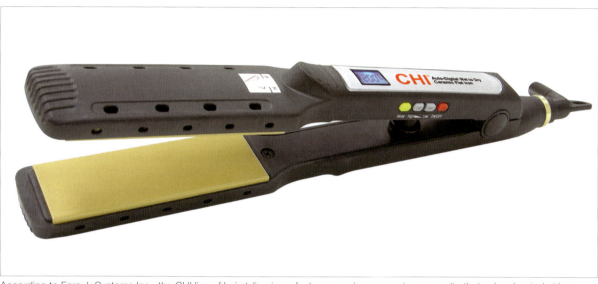

According to Farouk Systems Inc., the CHI line of hairstyling irons features a unique ceramic composite that, when heated at low temperatures, releases ions that are beneficial for hair, making it softer, smoother, and easier to style.

> *"Once you understand the mechanisms of the technology, you can look for spinoff applications."*

In addition, the incorporation of nanosilver led to the development of the Nano-CHI hairstyling and flat irons, Nano-hair dryers, and the CHI line of nail polish that incorporates nanosilver particles to inhibit microbial growth after the containers have been used.

After retiring from Johnson in 2006, Morrison started working with Shami as senior vice president of technology at Farouk Systems to use his NASA research experience as a platform to develop additional hair processing and styling tools that use near-infrared (NIR) light. Morrison's research into using light-emitting diode (LED) devices to improve skin healing and bone cell replacement in astronauts led to a new application for hair. He and Shami designed LED devices to apply the NIR light three times per week to stimulate scalp healing and the growth phase of hair follicles that have become dormant. Farouk Systems has been testing some CHI + NIR hair tools as well as complementary skin products, and aims to place them on the market in 2010.

According to Morrison, NASA expertise impacted the success of the CHI iron as well as opened the door for related developments. One of the latest is a digital controller that takes heat measurements and then adjusts the amount of current delivered several times per second to maintain an optimal temperature.

Morrison appreciates NASA's capability to do research and development without being focused on just one application. "Alternate uses may not be envisioned for a certain technology, but once you understand the mechanisms of the technology, you can look for spinoff applications," he explains. "As a NASA employee, I was encouraged to spread information about the concepts and results of our research, as well as talk to people about potential new applications of what we were discovering. The CHI hair iron is just a small example." ❖

CHI® is a registered trademark of Farouk Systems Inc.

These liquid formulations were specially designed to work with the combination of ions and infrared wavelengths created by Farouk Systems' CHI styling tools.

Do-It-Yourself Additives Recharge Auto Air Conditioning

Originating Technology/NASA Contribution

Even though it drops to -279 °F at night and dips to -400 °F inside its deepest craters, the Moon can reach a scorching 260 °F during the day. The range of temperatures is extreme—in part because there is no substantial atmosphere on the Moon to insulate against the heat or cold. What the Moon does have are small amounts of gasses above its surface, sometimes called a lunar atmosphere or exosphere, that consist mostly of hydrogen and helium, along with some neon and argon.

On Earth, traces of an atmosphere extend as high as 370 miles above the surface. Made of 78-percent nitrogen and 21-percent oxygen, 1 percent of Earth's atmosphere consists of argon and other gasses—some of which help to trap heat from the Sun and create a greenhouse effect. Without this effect, Earth would probably be too cold for life to exist. Another helpful feature of the Earth's atmosphere exists about 30 miles above the surface, where ultraviolet light from the Sun strikes oxygen molecules to create a gas called ozone. This ozone blocks harmful ultraviolet rays from reaching the Earth.

While the Earth's atmosphere protects and defends against extreme temperatures like those on the Moon, Earth's heating and air conditioning systems create an even more comfortable atmosphere indoors. In planning for a return mission to the Moon, NASA aimed to improve the thermal control systems that keep astronauts comfortable and cool while inside a spacecraft.

Partnership

In the late 1990s, Goddard Space Flight Center awarded a **Small Business Innovation Research (SBIR)** contract to Mainstream Engineering Corporation, of Rockledge, Florida, to develop a chemical/mechanical heat pump as part of the spacecraft's thermal control system. Designed to transfer heat from one location to another, a heat pump provides cooling by moving heat out of one area and into another. While working on the heat pump design at Goddard, Mainstream Engineering came up with a unique liquid additive called QwikBoost to enhance the performance of the advanced heat pump design.

Previously featured in *Spinoff* 1999, QwikBoost circulates through a system like a lubricant, working to boost the available cooling capacity. This increases the performance of the system and results in faster heat transfer (cooling) and consumption of less operating energy.

After Mainstream Engineering patented the QwikBoost technology developed with NASA, it started manufacturing and selling the additive to improve the operating efficiency and economy of refrigeration systems, air conditioners, and heat pumps. NASA used

In planning for a return mission to the Moon, NASA sought to improve the thermal control systems that keep astronauts comfortable while inside a spacecraft like the Lunar Module "Eagle," shown here on the far right.

> *"Working with NASA technology bolsters our confidence that the chemistry has been thoroughly tested and proven."*

QwikBoost to develop more efficient, smaller, and lighter cooling systems, as well as in air conditioning and refrigeration systems at NASA facilities, and in air conditioning systems in NASA's vehicle fleet.

Recognizing the capabilities of QwikBoost, a New York-based company, Interdynamics Inc., exclusively licensed the additive from Mainstream Engineering in 2004. As a developer of do-it-yourself air conditioning recharger kits, Interdynamics soon merged with EF Products Inc., of Dallas, Texas, a provider of closed system retrofit kits for automotive air conditioning systems, to become IDQ Inc., of Garland, Texas, with sales and marketing out of Tarrytown, New York. Today, IDQ incorporates the NASA-derived QwikBoost technology into its line of Arctic Freeze products.

According to the company, by using Arctic Freeze to replace lost refrigerant and oil in an automotive air conditioning system, the NASA-derived QwikBoost chemistry provides colder air up to 50-percent faster than a conventional R-134a refrigerant product. "Working with NASA technology bolsters our confidence that the chemistry has been thoroughly tested and proven to deliver the benefits and results promised," says Vincent Carrubba, director of research and development at IDQ.

Product Outcome

IDQ provides a variety of automotive air conditioning products for the do-it-yourself consumer and professional service technician, including its line of Arctic Freeze products. Sold at leading automotive and mass-retail stores and through wholesale distributors in the aftermarket industry in the United States, Europe, and Latin America, Arctic Freeze restores cooling in a vehicle's air conditioning system once the system is no longer cool-

IDQ Inc.'s Arctic Freeze-1 product recharges the air conditioning in most passenger automobiles manufactured after 1995. It comes with step-by-step instructions, a built-in reusable installation hose, snap-on coupler, and air conditioning pressure gauge.

The full line of Arctic Freeze products incorporates a QwikBoost refrigerant enhancer originally developed by NASA and Mainstream Engineering Corporation, of Rockledge, Florida. According to IDQ, QwikBoost provides vehicle owners with colder air up to 50-percent faster than a conventional refrigerant product.

ing effectively or when the performance has degraded to blowing only warm air. The product replenishes a system with R-134a containing the QwikBoost synthetic refrigerant enhancer.

Compared to operating with only PAG-oil (a lubricant), the addition of QwikBoost reduces wear and tear on the system by lowering compressor temperatures and extending the useful life of the lubricant. Arctic Freeze also incorporates a system-safe leak sealer that conditions rubber o-rings, seals and hoses, which are the primary source of minor system leaks.

In addition to delivering low vent temperatures, Arctic Freeze also delivers low costs. Depending on which Arctic Freeze product a customer uses, recharging an automotive air conditioning system can cost approximately $15–$30, compared to $100 or more at an automotive repair shop. Each Arctic Freeze product provides do-it-yourself customers with everything needed to recharge a vehicle air conditioning unit.

Carrubba believes NASA technology has made a world of difference by providing a demonstrable and affordable solution to improve the efficiency and economy of operating air conditioning and refrigeration systems here on Earth. "The all-in-one solutions of Arctic Freeze make it possible for nearly anyone to safely, effectively, and affordably recharge their own vehicle's air conditioning unit." ❖

QwikBoost™ is a trademark of Mainstream Engineering Corporation. Arctic Freeze® is a registered trademark of IDQ Inc.

Environmental Resources

While much of NASA's research is focused on expanding our understanding of the universe, some of the Agency's most important research is focused on our home planet, Earth. The technologies featured in this section:

- Analyze Water Quality in Real Time
- Expand Climate Knowledge
- Deliver Clean, Affordable Power
- Remediate Contaminated Groundwater
- Provide Cleanup of Oil Spills, Wastewater
- Protect People and Animals

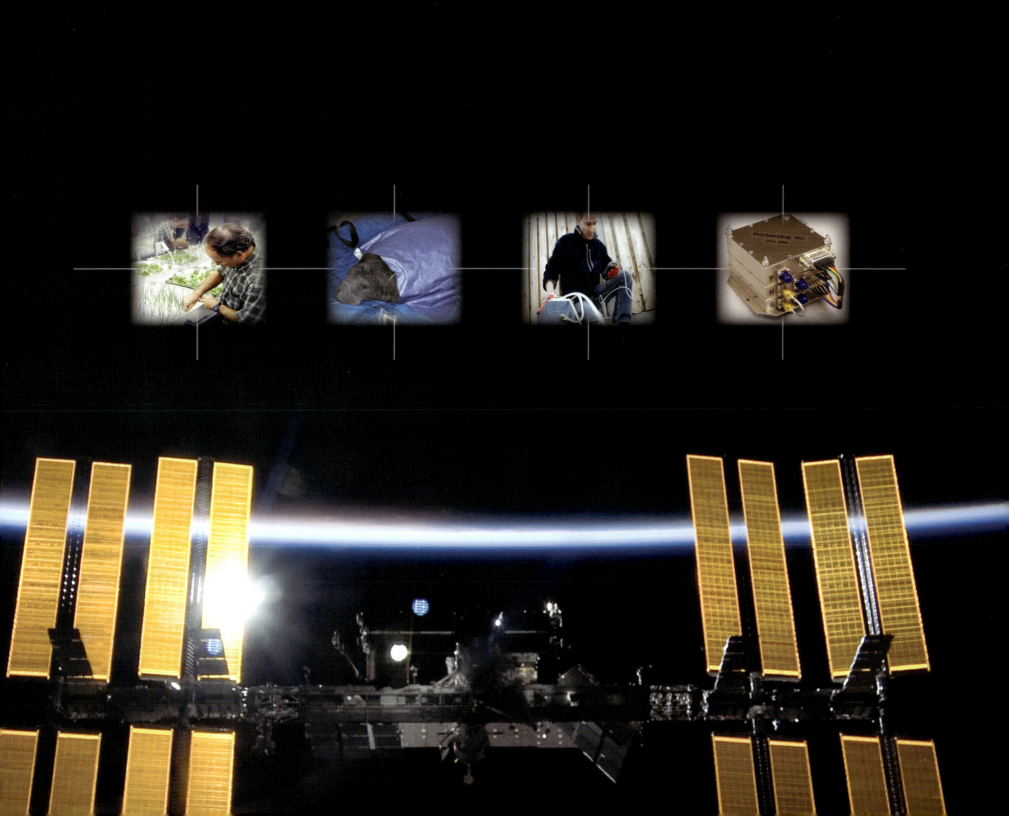

Systems Analyze Water Quality in Real Time

Originating Technology/NASA Contribution

If you are planning a day at your local park or even a weekend camping trip, you would simply pack all the food, drink, and supplies you need. Similarly, astronauts on short-term space missions can get away with packing the provisions they need to survive in space. But long-term space travel—a round-trip journey to Mars, for example—is no picnic. Depending on the mission, astronauts would need enough food to last for several months or years; a means to clean the air and water; and some efficient way of dealing with waste. Given cost and space limitations, packing supplies on this scale may not be feasible, meaning astronauts need a system that provides a steady source of sustenance while at the same time purifying and recycling waste. The best solution, then, for future deep space explorers: Develop a green thumb.

Earth is a closed, self-sustaining system. We depend on the planet's regenerative ecosystems to supply all of our biological needs. The central facilitators of these ecosystems are plants. The same qualities that make plant life essential on Earth make it an ideal engine for long-term life support in space. Plants actively purify and recycle air supplies, absorbing carbon dioxide and noxious fumes while releasing oxygen. They also provide a regenerating food supply, transpire water vapor into the air that can be condensed and collected for drinking, and filter contaminants in water through their root systems. For these reasons, NASA has long studied methods for growing and maintaining plants in space as part of a controlled (or closed) ecological life support system (CELSS) for long-term space missions.

Growing plants in space presents unique challenges. For one, soil may not be optimal due to its payload weight, the problem of particles floating about in zero gravity, and the potential for harboring health-threatening, soil-borne pathogens. To address this issue, many NASA projects—including the BioHome, the CELSS Breadboard Project,

Hydroponic techniques—like those used to grow these onions, lettuce, and radishes in this plant growth chamber in Kennedy Space Center's Space Life Sciences Laboratory—may one day contribute to the development of a bioregenerative life support system.

the Advanced Astroculture Spaceflight Experiments, and the Lunar-Mars Life Support Test Project—have focused on developing and improving hydroponic methods of plant cultivation for use in space. Hydroponics is the practice of growing plants in nutrient solutions, without true soil. This method not only proves more practical for space-based agriculture, but also typically produces larger, healthier plants.

Partnership

As part of NASA's Advanced Life Support Program, Kennedy Space Center sought methods for monitoring the nutrient components in hydroponic solutions. The solutions typically contain a variety of mineral elements and ions that are taken up at different rates by different plants and need to be replenished when their concentrations drop below ideal levels. Working with Kennedy scientist Dr. Ken Schlager under Phase I and II **Small Business Innovation Research (SBIR)** contracts, Biotronics Technologies Inc., of Waukesha, Wisconsin, developed a process analyzer that provides real-time detection of nutrients, organics, and metals in water.

Biotronics recognized a host of uses for the technology beyond hydroponic solution monitoring, including water and wastewater analysis and certain industrial

> *More than 500 ChemScan systems monitor treatment processes at water and wastewater facilities around the world.*

applications. Commercial funding from W. R. Grace & Co. led Biotronics to adapt the SBIR-developed technology for analyzing solutions related to Grace's cooling and boiler water treatment business. Biotronics then began exploring environmental applications, particularly water and wastewater monitoring at municipal treatment facilities. Realizing the commercial potential of the technology, a group of researchers at Biotronics purchased the rights to the technology from the company and created Applied Spectrometry Associates Inc. (now called ASA Analytics Inc.), also headquartered in Waukesha.

Product Outcome

The original process analyzer technology developed with NASA SBIR funding now takes the form of ASA's ChemScan product line. More than 500 ChemScan analyzer systems currently monitor the treatment processes at water and wastewater facilities around the world and help maintain water quality in major American cities like Los Angeles, New York, Phoenix, Las Vegas, Orlando, and Tampa.

"Water and wastewater treatment plants are basically biochemical factories," says Bernie Beemster, president of ASA. "Water is rarely found in pure form. It contains any number of contaminants and other chemical and biological constituents that need to be found and removed." The processes that cleanse wastewater and make tap water potable require a number of steps, Beemster says, each of which have control requirements based on measured parameters. Using fully automated basic spectrometry and an array-based absorbance spectrometer, the ChemScan analyzer measures absorbance levels across 256

The ChemScan analyzer ties directly into a treatment facility's control system, minimizing the need for operator intervention.

wavelengths of ultraviolet and visible light in a sample. The resulting data are processed using chemometrics—mathematical and statistical algorithms used in this case to determine an absorption pattern that indicates the presence of a particular chemical.

Earlier technology would require water samples to be physically extracted from various points along the treatment process and taken to a lab for analysis—a time consuming and labor intensive procedure that left significant gaps between measurements. ChemScan's capabilities allow it to automatically draw and analyze samples for multiple parameters within a water treatment process, including ammonia and other nutrient levels, water hardness, chloramination (a disinfectant process), or amounts of natural organic matter. The analyzer ties directly into the treatment facility's control system, which uses the information ChemScan provides to automatically adjust its processes, minimizing the need for operator intervention. ChemScan analyzers require little maintenance or calibration, making them a cost-efficient technology.

"The sum total of what ChemScan provides is a small fraction of the price that would be required to do the same task with other technologies," says Beemster.

While populations across the globe can credit the ChemScan analyzers with helping ensure the quality of their water supplies, Beemster credits the NASA SBIR-developed technology as the core reason for his company's success. "Over the 15 years of our history, we've had a compound annual growth rate of 40 percent as a direct result of this unique technology," Beemster says. "That's pretty remarkable." ❖

ChemScan® is a registered trademark of ASA Analytics Inc.

Compact Radiometers Expand Climate Knowledge

Originating Technology/NASA Contribution

NASA not only peers up to gather information about space; it also peers down to gather information about Earth. As part of the Science Mission Directorate, NASA's Earth Science Program aims to improve predictions about climate, weather, and natural hazards by understanding Earth's response to natural and human-induced changes. One way scientists are tracking these changes is by monitoring the Earth's soil moisture and ocean salinity.

As part of the water cycle, soil moisture and ocean salinity are interconnected with Earth's energy and biogeochemical cycles (carbon, nitrogen, and others). Together, these natural cycles play a role in moderating the overall environment. For example, ice can melt into the ocean to dilute ocean salinity, which affects ocean circulation, which affects an area's climate.

In 2015, NASA plans to embark on a mission called Soil Moisture Active and Passive (SMAP) to provide global measurements of soil moisture and its freeze and thaw states. Measurements from the SMAP spacecraft will lead to a better understanding of the processes that link the water, energy, and carbon cycles, and will significantly inform Earth system science, water resource assessment, and natural hazards mitigation.

The best way to measure surface conditions such as soil moisture and ocean salinity is through microwave remote sensing at long wavelengths, such as at L-band. However, receiving long wavelengths from Earth requires large antennas in space. To prepare for microwave remote sensing at long wavelengths, NASA has supported the development of smaller, lighter, more compact devices to uncover fluctuations in soil moisture and ocean salinity.

Partnership

As a result of **Small Business Innovation Research (SBIR)** funding from Goddard Space Flight Center in 2000, ProSensing Inc., of Amherst, Massachusetts, developed a compact, ultrastable radiometer for sea surface salinity and soil moisture mapping. Taking advantage of the rapid advances in telecommunications at the time, ProSensing incorporated small, low-cost, high-performance elements into just a few circuit boards. By using two or more radiometers (called arraying) and combining the signals they receive, the radiometers act as small pieces of one large system.

"It was important to make them small enough to hold in the palm of your hand. Previous radiometers were about 10 times the size. If we built radiometers the old way, it would take a lot more time, money, and effort. By using cell phone technology, it cost 50-percent less

NASA's Soil Moisture Active and Passive mission, planned for 2015, will gather global measurements of soil moisture and its freeze and thaw states to inform the scientific community's understanding of climate, the carbon cycle, and water resources.

and gave us a competitive advantage," says James Mead, president of ProSensing.

By 2005, ProSensing had delivered 35 units to Goddard for use in a research instrument called 2D-STAR (Two Dimensional Steered Thin Array Radiometer). Flown on an aircraft, 2D-STAR was developed to demonstrate interferometric technology (arraying radiometers to act as one large radiometer) for remote sensing of soil moisture.

David M. Le Vine, a scientist at Goddard Space Flight Center, says, "2D-STAR achieves the results of a large antenna without putting a massive structure in space. We were able to show the technology could work from airplanes, with the idea of going from airplanes to space."

Product Outcome

During its work with Goddard, ProSensing received a request from the University of Melbourne, Australia, to build six ultrastable radiometer modules for airborne remote sensing at a new facility being developed through the support of the Australian Research Council, University of Melbourne, University of Newcastle, James Cook University, and Airborne Research Australia at Flinders University. By 2006, ProSensing had built and delivered a polarimetric L-band microwave radiometer (PLMR) using the radiometer modules.

According to the University of Melbourne, PLMR enables mapping in unprecedented detail of soil moisture, land surface salinity and temperature, and ocean surface salinity and temperature. The University appreciates the instrument's small size, light weight, and affordability.

Based on the advances made possible by working with NASA, ProSensing offers two lightweight radiometers for mapping soil moisture, ocean salinity, and wind speed: the PLMR and the Scanning Low Frequency Microwave Radiometer. ProSensing's customers include government research agencies, university research groups, and large corporations in North America, Europe, and Asia.

The way a microwave radiometer works is by measuring natural radiation. In the case of the ocean, the amount of signal that is radiated changes with salinity. The parameter used to describe this is called the emissivity of the surface. In the case of soil, the emissivity changes with moisture. Additionally, the emissivity depends on how rough a surface is, and this is why a radiometer can also measure wind speed over the ocean. The faster the wind blows, the rougher the surface gets, resulting in more signals being transmitted to the radiometer.

In 2008, the NASA-derived technology influenced the design and development of a hurricane imaging radiometer (HIRAD) to measure ocean wind speed in hurricanes. The instrument, under development at Marshall Space

ProSensing Inc.'s ultrastable compact radiometers are employed around the world to map soil moisture, ocean salinity, and wind speed.

Flight Center, uses the same technique as the ultrastable radiometer, but operates at a different wavelength.

HIRAD is a compact, lightweight, low-power instrument that produces imagery from an aircraft or spacecraft. Specifically, Marshall is developing HIRAD to measure ocean surface wind speed from hurricane reconnaissance aircraft with fine spatial resolution over a wide angular swath. ProSensing delivered 12 single-board, four channel radiometer receivers to Marshall, which are now being integrated into HIRAD.

In 2009, ProSensing received a contract from one of the largest research centers in Germany, the Institute of Chemistry and Dynamics of the Geosphere, to build a copy of the PLMR. Requiring six ultrastable radiometer modules, the Institute uses PLMR for research measurements of soil moisture and salinity.

Most recently, the University of Massachusetts requested two radiometer modules from ProSensing to

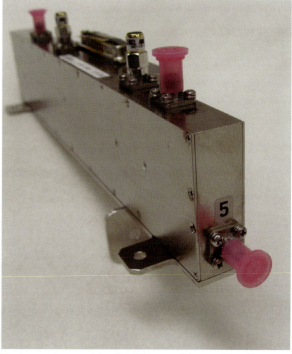

Twelve receiver modules like this one will be combined in a hurricane imaging radiometer to measure wind speeds in hurricanes.

test their applicability for noninvasive readings of cerebrospinal fluid temperature (temperature of the fluid that circulates around and through the brain and spinal cord). Usually, a spinal tap is used for this procedure to diagnose certain neurologic disorders, particularly infections like meningitis. "If you can take temperature remotely with a radiometer and see that it is elevated, then this could have wide application in emergency rooms," says Mead.

Mead says many projects have grown out of the NASA SBIR program over the years. "NASA had a long-term interest in developing the technology and was eager to take what we built and put it to use. That gave us something to design to. Now we have some fairly mature projects, and some that are just getting kicked off." ❖

Energy Servers Deliver Clean, Affordable Power

Originating Technology/NASA Contribution

Imagine you are about to be dropped in the middle of a remote, inhospitable region—say the Kalahari Desert. What would you want to have with you on your journey back to civilization? Food and water, of course, but you can only carry so much. A truck would help, but what would you do when it runs out of gas? Any useful resources would have to be portable and—ideally—sustainable.

Each of Bloom Energy's fuel cells, such as those held by company CEO K.R. Sridhar above, can produce 25 watts—enough to power a light bulb. The cells can be stacked to increase the power output. The lime-green coating is the anode side; the black coating is the cathode side.

Astronauts on future long-term missions would face similar circumstances as those in this survivalist scenario. Consider, for example, a manned mission to explore the surface of Mars. Given the extreme distance of the journey, the high cost of carrying cargo into space, and the impossibility of restocking any supplies that run out, astronauts on Mars would need to be able to "live off the land" as much as possible—a daunting proposition given the harsh, barren Martian landscape. Not to mention the lack of breathable air. Another consideration is fuel; spacecraft might have enough fuel to get to Mars, but not enough to return. The Moon is like a day trip on one tank of gas, but Mars is a considerably greater distance.

In the course of planning and preparing for space missions, NASA engineers consistently run up against unprecedented challenges like these. Finding solutions to these challenges often requires the development of entirely new technologies. A number of these innovations—inspired by the extreme demands of the space environment—prove to be solutions for terrestrial challenges, as well. While developing a method for NASA to produce oxygen and fuel on Mars, one engineer realized the potential for the technology to generate something in high demand on Earth: energy.

Partnership

K.R. Sridhar was director of the Space Technologies Laboratory at the University of Arizona when Ames Research Center asked him to develop a solution for helping sustain life on Mars. Sridhar's team created a fuel cell device that could use solar power to split Martian water into oxygen for breathing and hydrogen for use as fuel for vehicles. Sridhar saw potential for another application, though. When the NASA Mars project ended in 2001, Sridhar's team shifted focus to develop a commercial venture exploring the possibility of using its NASA-derived technology in reverse—creating electricity from oxygen and fuel.

On the surface, this sounds like standard hydrocarbon fuel cell technology, in which oxygen and a hydrocarbon fuel such as methanol flow into the cell where an electrolyte triggers an electrochemical reaction, producing water, carbon dioxide, and electrons. Fuel cells have provided tantalizing potential for a clean, alternative energy source since the first device was invented in 1839, and NASA has used fuel cells in nearly every mission since the 1960s. But conventional fuel cell technology features expensive, complicated systems requiring precious metals like platinum as a catalyst for the energy-producing reaction. Sridhar's group believed it had emerged from its NASA work with innovations that, with further development, could result in an efficient, affordable fuel cell capable of supplying clean energy wherever it is needed.

In 2001, Sridhar's team founded Ion America and opened research and development offices on the campus of the NASA Research Park at Ames Research Center. There, with financial backing from investors who provided early funding to companies like Google, Genentech, Segway, and Amazon.com, the technology progressed and began attracting attention. In 2006, the company delivered a 5-kilowatt (kW) fuel cell system to The Sim Center, a national center for computational engineering, at The University of Tennessee Chattanooga, where the technology was successfully demonstrated. Now called Bloom Energy and headquartered in Sunnyvale, California, the company this year officially unveiled its NASA-inspired technology to worldwide media fanfare.

Product Outcome

Bloom Energy's ES-5000 Energy Server employs the planar solid oxide fuel cell technology Sridhar's team originally created for the NASA Mars project. At the core of the server are square ceramic fuel cells about the size of old fashioned computer floppy disks. Crafted from an inexpensive sand-like powder, each square is coated with special inks (lime-green ink on the anode side, black on the cathode side) and is capable of producing

> *"NASA is a tremendous environment for encouraging innovation. It's all about solving problems that are seemingly unsolvable."*

25 watts—enough to power a light bulb. Stacking the cells—with cheap metal alloy squares in between to serve as the electrolyte catalyst—increases the energy output: a stack about the size of a loaf of bread can power an average home, and a full-size Energy Server with the footprint of a parking space can produce 100 kW, enough to power a 30,000-square-foot office building or 100 average U.S. homes.

Solid oxide fuel cells like those in Bloom's Energy Server operate at temperatures upwards of 1,800 °F. The high temperatures, efficiently harnessed by the Bloom system's materials and design, enable the server to use natural gas, any number of environmentally friendly biogasses created from plant waste, or methane recaptured from landfills and farms. Fuel is fed into the system along with water. The high temperatures generate steam, which mixes with the fuel to create a reformed fuel called syngas on the surface of the cell. As the syngas moves across the anode, it draws oxygen ions from the cathode, and an electrochemical reaction results in electricity, water, and only a small amount of carbon dioxide—a process that according to Bloom is about 67-percent cleaner than that of a typical coal-fired power plant when using fossil fuels and 100-percent cleaner with renewable fuels. The server can switch between fuels on the fly and does not require an external chemical reformer or the expensive precious metals, corrosive acids, or molten materials required by other conventional fuel cell systems.

The technology's "plug and play" modular architecture allows users to generate more power by simply adding more servers, resulting in a "pay as you grow" scenario in which customers can increase their energy output as their needs increase. The Bloom Energy Server also offers the benefits of localized power generation; the servers are located on site and off the grid, providing full-time power—as opposed to intermittent sources like solar and wind—without the inefficiencies of transmission and distribution, Bloom says. Future servers may even return to the original NASA function of using electricity to generate oxygen and hydrogen. The company envisions feeding electricity from wind or solar power into its servers along with water to produce storable hydrogen and oxygen. The server could then use the stored gasses to generate electricity during cloudy, low-wind, and nighttime conditions. Stored hydrogen could even be used to provide fuel for hydrogen-powered cars.

Bloom quietly installed its first commercial Energy Server in 2008, and since then its servers have generated more than 11 million kilowatt hours (kWh) of electricity, along with a corresponding 14-million-pound reduction in carbon dioxide emissions, which the company says is the equivalent of powering about 1,000 American homes for 1 year and planting 1 million trees. Bloom's current customers are a who's-who of Fortune 500 companies, including Google, eBay, Bank of America, The Coca-Cola Company, and FedEx. Bloom says its customers can expect a return on their investment from energy cost savings within 3–5 years, and eBay has already claimed more than $100,000 in savings on electricity expenses.

Sridhar believes it will be another 5 to 10 years before Bloom's technology becomes cost-effective for home use. At that point, he sees the Bloom Energy Server as a solution for remote and underdeveloped areas in need of power. He says the company's mission is "to make clean, reliable energy affordable to everyone in the world."

"One in three humans lives without power," Sridhar says. "Energy demand exceeds supply." Just within the United States, 281 gigawatts of new generating capacity—the output of 937 new 300-megawatt power plants—will be necessary by 2025 to meet national energy demands, according to the U.S. Energy Information Administration.

The Bloom Energy Server may soon offer an environmentally sound option for meeting that challenge, a solution derived from the demands of space exploration.

"NASA is a tremendous environment for encouraging innovation," says Sridhar. "It's all about solving problems that are seemingly unsolvable. After realizing we could make oxygen on Mars, making electrons on Earth seemed far less daunting. We're grateful to NASA for giving us a challenge with serendipitous impact for mankind." ❖

Energy Server™ is a trademark of Bloom Energy.

Bloom Energy Servers, seen here on eBay's corporate campus, are now providing environmentally friendly, cost-saving energy to a number of Fortune 500 companies.

Solutions Remediate Contaminated Groundwater

Originating Technology/NASA Contribution

Kennedy Space Center's launch complexes have seen a lot. They have been the starting point for every manned NASA mission, from Mercury to Gemini, through Apollo, and are now seeing the space shuttle through its final launches. Kennedy, part of the Merritt Island Wildlife Refuge, is also home to over 1,500 different plant and animal species.

To help protect these living things, NASA works to keep the area as pristine as possible and that sometimes involves inventing new and innovative ways to clean up around the launch pads.

During the Apollo Program, NASA launched Saturn rockets from Launch Complex 34 (LC-34), and during that period—roughly 1959 to 1968—workers used chlorinated solvents to clean rocket engine components. An estimated 88,000 pounds of these chlorinated solvents, which are only slightly soluble and heavier than water, soaked into the soil and groundwater. Even after several decades, the contaminants were still in the groundwater and would have stayed even longer (restoration of the area through natural processes would have taken many decades), had NASA not set about searching for ways to remediate the area.

Partnership

These solvents belong to a category of chemicals known as dense non-aqueous phase liquids (DNAPLs). DNAPLs had contaminated the areas around LC-34 to the point of near-irreparability. NASA environmental engineer Dr. Jacqueline Quinn and Dr. Kathleen Brooks Loftin, a NASA analytical chemist, partnered with researchers from the University of Central Florida's chemistry and engineering programs to develop an innovative technology capable of remediating the area without great cost or further environmental damage. They called the new invention Emulsified Zero-Valent Iron (EZVI). It neutralizes the toxic chemicals using nano- or micro-sized iron particles in an environmentally friendly water and biodegradable oil emulsification. EZVI is injected deep into the soil, where the contaminants diffuse through the emulsion's oil membrane and are then dechlorinated by the iron-water interior. The only byproduct of this process is non-toxic hydrocarbon, which diffuses into the groundwater.

EZVI was field-tested by the U.S. Environmental Protection Agency under the Superfund Innovative Technology Evaluation Program and is now used at commercial and government sites to treat heavily contaminated areas.

Once the NASA-University team had EZVI working in the laboratory, it then contracted with GeoSyntec Inc., a small environmental remediation firm in Boca Raton, Florida, through the **Small Business Technology Transfer (STTR)** program. GeoSyntec worked at LC-34 to evaluate the real-world effectiveness of EZVI, which had previously only been laboratory tested. Groundwater

> The NASA-developed EZVI groundwater remediation compound is NASA's most-licensed technology to date.

testing showed that EZVI was an effective means of removing the DNAPLs. Through a follow-on contract, GeoSyntec began testing various other methods for injecting EZVI into the soil.

EZVI was recognized as a 2005 "NASA Government Invention of the Year" and 2005 "NASA Commercial Invention of the Year." In 2006, the inventors won the Federal Laboratory Consortium's "Award for Excellence in Technology Transfer." In 2007, EZVI was inducted into the Space Foundation's prestigious Space Technology Hall of Fame. According to Kevin Cook, director of Space Awareness Programs at the Space Foundation, "EZVI is an environmentally friendly technology that effectively decontaminates polluted soil and groundwater and is an outstanding example of inspired innovations we seek to induct into the Space Technology Hall of Fame."

Since its development, numerous companies have licensed use of this technology from NASA. Several licenses are in the works, but currently six companies are using the NASA-developed EZVI groundwater remediation compound to clean up polluted areas all around the world, making it NASA's most-licensed technology to date.

Licensees include Weston Solutions Inc., of Westchester, Pennsylvania, which provides environmental remediation services worldwide; and Toxicological and Environmental Associates, Inc., of Baton Rouge, Louisiana, which is offering the services through the southern and central United States (*Spinoff* 2005). Additional licensees include Huff and Huff Inc., of Oak Brook, Illinois, which offers EZVI remediation as one of its many environmental cleanup options; Starlight Environmental Group, of Panama City Beach, Florida; Remediation and Natural Attenuation Services Inc. (RNAS), of Brooklyn Center, Minnesota; and Terra Systems Inc., of Wilmington, Delaware.

Product Outcome

The EPA estimates that there are thousands of sites around the United States contaminated with DNAPLs, including 60 to 70 percent of the sites on the Superfund National Priorities List, a comprehensive list of abandoned hazardous waste sites receiving top priority for environmental remediation. As such, EZVI was accepted into the EPA's Superfund Innovative Technology Evaluation Program, which promotes the advancements of innovative technologies for detection and remediation of contaminated superfund-designated sites.

EZVI has proven a valuable tool in remediating these sites, and has also been used at dye, chemical, pharmaceutical, adhesive, aerosol, and paint manufacturing sites; dry cleaners; leather tanning facilities; metal cleaning and degreasing facilities; and many government-owned sites. The licensees of NASA's EZVI technology are cleaning these contaminants from groundwater all over the country, in a wide variety of soil types, including sand and silt, clay, fractured bedrock, and residuum (soil formed in place by natural weathering).

The EZVI treatment methods generally take between 2 and 3 months, a vast improvement over traditional pump-and-treat systems, which can require decades. This reduced timeframe, in addition to the low cost of the materials and equipment, make EZVI cost-competitive with pump-and-treat, thermal treatments, and in situ chemical oxidation. In addition, some of these other methods have the potential to release the DNAPLs to previously uncontaminated areas. Since EZVI treats the contaminants in place, there is no risk for them to mobilize. ❖

EZVI directly treats the contaminant source without mobilizing contaminants. The environmentally safe treatment method is cost-competitive, requires less treatment time than traditional methods, and produces less toxic, more easily biodegradable byproducts.

Bacteria Provide Cleanup of Oil Spills, Wastewater

Originating Technology/NASA Contribution

Given the size of our planet and its wealth of resources, it is easy to forget that those resources are finite. As Earth's human population continues to grow, the questions of how to effectively limit and recycle waste, avoid environmental contamination, and make the most of water and fuel reserves become all the more pressing.

On a much smaller scale, these same concerns apply to astronauts living within the closed system of the International Space Station (ISS). All resources onboard the ISS—air, water, energy—are limited and must be carefully managed and recycled to create a sustainable environment for the crewmembers. This challenge must be met without the natural systems that provide for and sustain life on Earth.

Well before construction of the ISS began in 1998, NASA was investigating ways water could be purified and reused by astronauts living in orbit. One method the Agency explored—through partnership with a small Texas company—involved bringing into space Earth's most abundant biological resource: bacteria.

Partnership

Micro-Bac International Inc., headquartered in Round Rock, began business in the early 1980s with the idea to selectively utilize Earth's natural waste management system to provide safe, efficient, and environmentally sound solutions for a host of applications.

"In the biosphere, everything gets broken down by microorganisms," says Dennis Schneider, vice president and director of research and development for Micro-Bac. "But in specific waste applications, you find that the right mix of microorganisms is not there. What we've found over the years is that we can isolate microorganisms out of the environment, and individual strains of those would have the capacity to break down certain types of organic compounds that are typically difficult to degrade." The bacteria accomplish this, Schneider explains, by producing protein enzymes that break down organic compounds into subunits, which the bacteria then grow on, creating more bacteria, carbon dioxide, and water.

"It's a natural process with no toxic byproducts," says Schneider.

Through Phase I and II **Small Business Innovation Research (SBIR)** contracts with Marshall Space Flight Center, the company developed a phototrophic cell for water purification. Inside the cell: millions of photosynthetic bacteria from strains specifically isolated for their ability to break down toxic chemicals astronauts could encounter on the ISS. Requiring only enough light to sustain the bacteria, the cell could provide a low-power option for cleansing wastewater during long-term space missions.

Micro-Bac proceeded to commercialize the bacterial formulation it developed for the SBIR project. Mega-BacTF, first featured in *Spinoff* 1999, is among the microbial products the company offers to the benefit of cities and industry around the world.

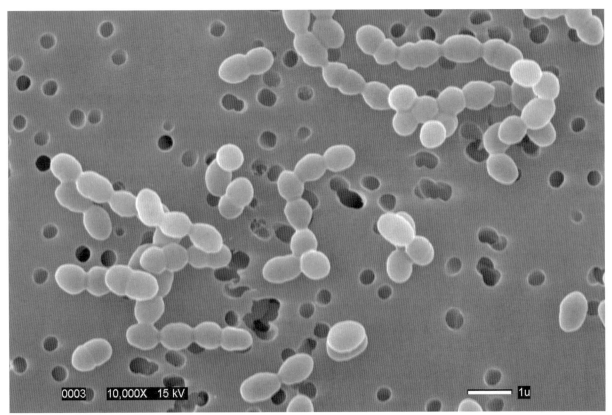

Micro-Bac International Inc.'s microbial solutions, including formulations developed under NASA SBIR contracts, utilize specifically selected bacteria combinations to naturally break down organic compounds such as animal waste and oil, without yielding toxic byproducts.

Product Outcome

Mega-BacTF is now part of an expanded Mega-Bac product line formulated for organic materials degradation and odor control in large bodies of water like municipal lagoons. Mega-Bac products are also used for the remediation of animal waste, wastewater systems, and septic tanks, and are employed in waste treatment for livestock farms and food manufacturers. The leading U.S. pork producer, tortilla plants, juice makers, microbreweries, and even tequilerias in Mexico use Micro-Bac's natural, nonpathogenic biotechnology to help limit the environmental impact of their waste byproducts. The company's bacterial solutions are also popular in tropical regions such as Brazil, where plentiful sunlight makes Micro-Bac's photosynthetic bacteria a cost-effective alternative to the traditional sludge systems used in municipal wastewater treatment. Micro-Bac also offers products designed to treat hazardous and contaminated waste; dairy waste; grease, fats, and oils; waste from fruit and vegetable processing; and waste from leather tanning.

"We're into just about anything you can imagine that involves organic material degradation," Schneider says. The company has collected bacterial species from around the world and carefully formulates its products using specific strains that work in harmony to target each customer's environmental issue—no genetic engineering involved and no special handling measures or equipment required.

"Wood, for example, is slow to break down in nature," Schneider says. "But you can find microorganisms that can break down wood very quickly. You can target specific applications."

Micro-Bac offers more than specific waste treatment solutions. The company is also a leading provider of microbial products for improving oil production. Oil wells often struggle with the accumulation of compounds like paraffins and asphaltenes, components of crude oil that can settle out and create deposits that can bring oil production to a standstill and require costly maintenance treatments to remove. Micro-Bac's oilfield products, which include strains of bacteria from the company's NASA-derived formulation, break down those deposits, leading to significant boosts in productivity—one Kansas oil well increased its production over 500 percent after treatment with Micro-Bac products.

These same qualities make Micro-Bac's oil-targeted products effective tools for countering environmentally damaging oil spills. The company has assisted in the cleanup of crude oil spills in Ecuador, and officials called upon Micro-Bac's microbial solutions to help mitigate the environmental impact of the catastrophic 2010 oil rig explosion off the coast of Louisiana by breaking down oil that reached shore.

Micro-Bac's products have been used for oil spill cleanup, such as at this site in the Amazon rainforest in Ecuador, and is being employed in mitigating the environmental damage caused by the 2010 oil rig explosion and spill in the Gulf of Mexico.

Micro-Bac continues to develop targeted biotechnology for its customers' waste treatment and oil production applications, including a granulated version of its liquid crude oil degrading products. Customers will be able to keep the dried product on hand to simply sprinkle on any small oil spills. Meanwhile, the derivatives of the company's small-scale NASA collaboration continue to help address large-scale environmental and energy concerns in the United States and beyond.

"A wellspring of utility has come out of that work," says Schneider. ❖

Mega-BacTF™ is a trademark of Micro-Bac International Inc.
Mega-Bac® is a registered trademark of Micro-Bac International Inc.

Reflective Coatings Protect People and Animals

Originating Technology/NASA Contribution

In the course of time, circumstance, and coincidence, life sometimes completes some ironic and unlikely circles. During NASA's early space exploration, a deceptively simple concept allowed scientists and engineers to manage thermal gain. They used highly reflective coated surfaces on ultra-light substrates to reflect infrared energy to effectively and reliably keep operating temperatures in the extremes of space at safe and manageable levels. That reflective insulation technology has kept thousands of satellites; all of NASA's manned spacecraft, the Hubble Telescope, and most importantly, astronauts, safe and functional. NASA engineers used this technology to reflect infrared energy the way a child learns to redirect a sunbeam with a hand-held mirror.

Seamstresses work on a sunshade for Skylab, which lost a thermal protection shield during its launch in 1973. The sunshade utilized a radiant barrier technology consisting of a thin plastic material coated with vaporized aluminum.

Fast forward to 2009, when a young marine biologist working among the waterways of Florida's East Coast, almost in the shadows of the iconic launch pad gantry sites at the Kennedy Space Center, employs that same NASA-developed principle that protects astronauts in space to manage the core temperature of one of nature's most primitive and endangered creatures: the gentle manatee. The chain of events that brought marine biologist Artie Wong and his colleagues to the ideal solution started almost four decades ago, adjacent to these same waterways.

On May 14, 1973, Skylab launched into orbit and became a stunning example of what was possible for the Space Program. As the name implies, Skylab served as a space-based laboratory as well as an Earth-observing facility and a home base for three crews of astronauts. While Skylab proved that humans could live and work in outer space for extended periods of time, it also proved that great achievements often come with great challenges.

During launch, one of the protective shields on the outside of the spacecraft loosened, causing one solar panel to fall off, and preventing another from deploying. To plan for the spacecraft's repair, Marshall Space Flight Center led a collaborative effort to start working around the clock.

Partnership

Over the next 10 days, NASA engineers called upon National Metalizing, of Cranbury, New Jersey, to help create a reflective parasol-type sunshield to deploy on Skylab in place of the protective shield. National Metalizing was a manufacturer of reflective material utilizing a radiant barrier technology originally developed for NASA in the 1950s. In the public domain for the last 30 years, the radiant barrier technology consists of a thin plastic material coated with vaporized aluminum to either deflect or conserve infrared energy.

Eleven days after Skylab launched, astronauts launched from a second spacecraft, visited, and deployed a parasol sunshade made with the radiant barrier technology. As for the maker of the material that saved the spacecraft, National Metalizing's manufacturing capacity was used for industrial coating and lamination applications after the company was sold in the mid-1980s.

In 1980, a former director of sales and marketing for the company, David Deigan, founded a new company recently renamed Advanced Flexible Materials (AFM) Inc. Headquartered in Petaluma, California, the company employs the same reflective insulation technology to produce ultra-light, compact travel and stadium blankets; colorfully printed wraps to keep hundreds of thousands of marathon finishers safe from hypothermia each year; a successful line of Heatsheets outdoor products sold in retail stores around the world under the Adventure Medical Kits label; as well as reflective insulating lining fabrics for mittens, vests, and more. In 1996, the radiant barrier technology was inducted into the Space Foundation's prestigious Space Technology Hall of Fame.

Product Outcome

Deigan began providing thin plastic blankets made with the radiant barrier technology to keep thousands of runners warm at the 1980 New York City Marathon. Depending on when and where a marathon takes place, temperatures can be cool, and when participants stop running, hypothermia can become a problem. Today, the Heatsheets for running events and triathlons are produced exclusively in an environmentally friendly, recyclable polyethylene form, and reflect up to 97 percent of a person's radiant heat, providing an envelope of warm air around the body to prevent hypothermia.

For nearly two decades, Alice and Bill Wong volunteered at the New York City Marathon, helping to distribute thousands of Heatsheets at the finish line. In 2009, Alice contacted Deigan to see if his company would be willing to donate some of its heat-reflective products for an unusual cause. Her son, Artie, a marine biologist with a non profit club called Save the Manatee Club was

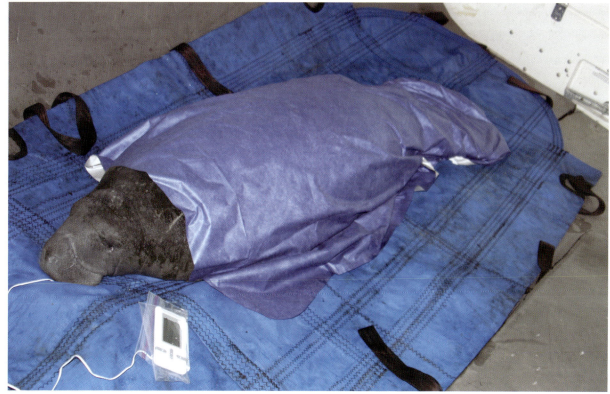

As part of a tag-and-release program, Advanced Flexible Materials Inc. (AFM) donated insulation with NASA-derived radiant barrier technology to keep manatees warm after being lifted from the water. A thermometer in the manatee's mouth monitors its temperature.

AFM's newest line of travel blankets, including a waterproof version and a breathable version, comes complete with a portfolio case.

searching for a solution to a unique challenge. The club's members work with the State of Florida Fish and Wildlife Research Institute to track and document the migration routes of manatees through Florida's waterways, some of them not far from the Kennedy launch pad gantries on Florida's East Coast.

As a part of the tag-and-release program, the docile sea cows are lifted from the water, checked to record their vital signs, tagged, and then returned to the water. Because these sea mammals are unable to withstand temperatures below 60 °F without slipping into hypothermia, Artie was searching for a more effective way to cover the manatees and conserve their vital body heat. The NASA-derived technology that gave his parents a way to help warm up runners for so many years seemed like the ideal solution. AFM recommended a more suitable solution, and donated a composite material incorporating a soft, non-woven fabric laminated with an encapsulated reflective insulation layer that could withstand the corrosive effects of salt water and could be safely reused for extended periods of time.

The finished version of the product has warmed more than 2 million Japanese consumers as an ultra-light, compact blanket; a futon cover; and a travel blanket. Over the past decade, AFM has exported to Japan through its trading partner, D. Nagata Co. Limited.

From helping to save Skylab to helping to save the manatees, the NASA-derived technology is sure to continue making a difference in the future, both on Earth and in space. ❖

Heatsheets® is a registered trademark of Advanced Flexible Materials Inc.

Computer Technology

NASA's computer scientists provide engineers with the computing resources and simulation tools to carry out critical NASA missions, but these same abilities translate into terrestrial benefits. The technologies featured in this section:

- Simplify Vibration Analysis
- Predict Flow in Fluid Dynamics
- Secure Online Shopping, Banking
- Maintain Health of Complex Systems
- Multiply Computing Power
- Automate Spacecraft Testing, Operation
- Deliver Positioning Solutions
- Enhance Scientific Data Collection
- Simulate Fine Particle Dispersion

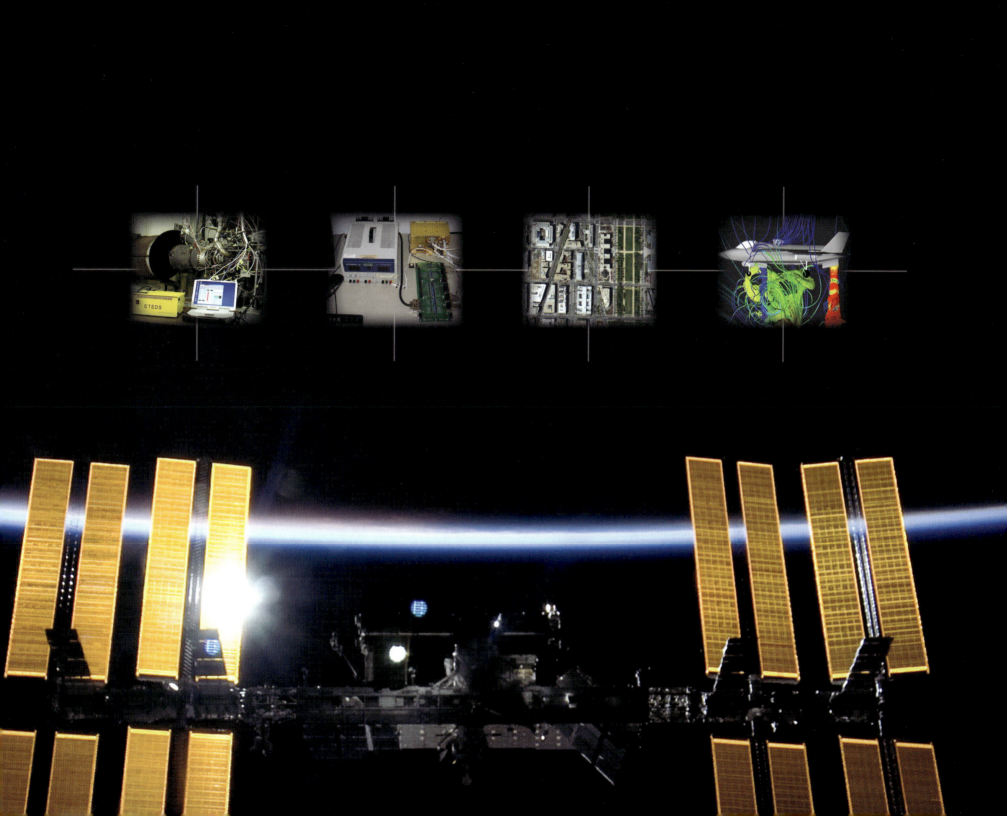

Innovative Techniques Simplify Vibration Analysis

Originating Technology/NASA Contribution

As the launch clock counts down, astronauts in the space shuttle prepare for the fastest ride of their lives. More powerful than any plane, train, or automobile, NASA space shuttles boast the world's most sophisticated rocket engines: three 14-foot-long main engines that produce more than 375,000 pounds of thrust each. This thrust is approximately four times that of the largest commercial jet engine—and produces an extreme amount of vibration.

In the early years of development, Marshall Space Flight Center engineers encountered challenges related to the development of engine components in the space shuttle main engine (SSME). To assess the nature of these problems, the engineers evaluated the system's components and operating environment, including the effects of vibration and oscillation. Similar to examining a patient's heartbeat and blood pressure to inform a medical diagnosis, engineers measure vibrations to diagnose mechanical issues. At the time, vibration signal analysis was a costly, tedious, and time-consuming process. It required an expansive in-house computer system along with several dedicated personnel to keep it functioning.

"A number of significant hardware failures on the SSME necessitated the need for more sophisticated and advanced diagnostic analysis techniques to be developed," says Jess Jones, a retired NASA engineer and a senior staff engineer at AI Signal Research, Inc. (ASRI) in Huntsville, Alabama.

Partnership

By the 1990s, the power of personal computers advanced significantly, and Marshall awarded ASRI a **Small Business Innovation Research (SBIR)** contract to enhance the method of vibration signal analysis. Through follow-on SBIRs with Marshall, ASRI developed PC-SIGNAL, a software package with improved tools and techniques for vibration signal analysis to assess engine ground test and flight performance. For the first time, engineers could analyze vibration and oscillation data from the convenience of a laptop or personal computer.

The new techniques allowed for quick and easy identification of potential design issues related not only to vibration, but also related to other signals resulting from sound, strain, and fluid flow. Even though the amount of sophisticated data increased, PC-SIGNAL simplified the analysis of it. According to Dr. Jen Jong, the chairman and director of research and development at ASRI, "Signal analysis had been limited to time, frequency, and phase domains, but PC-SIGNAL offered advanced

Space Shuttle Endeavour concludes a mission at Dryden Flight Research Center in 1992. The three liquid-propelled main engines, seen at the back of the vehicle, help to power the shuttle into low Earth orbit.

> *"Partnering with NASA provided a unique opportunity to work on challenging technical problems."*

signal analysis at various other domains. This information provided further insight into the engine's operating conditions."

One of PC-SIGNAL's unique modules, PKP-Module, reconstructs speed signals directly from external vibration measurements, without the installation of intrusive speed sensors. This enables additional vibration analysis techniques for the SSME.

"Partnering with NASA provided a unique opportunity to work on challenging technical problems, which in turn stimulated research and development of innovative solutions to real-world problems. It allowed ASRI to participate in the advancement of state-of-the-art techniques in many critical engineering areas," says Jong.

Today, NASA employs PC-SIGNAL on a daily basis for development and testing of propulsion systems. It is used for processing dynamic data and other diagnostic analysis, such as analyzing engine health data from shuttle missions, flow testing to identify and resolve fluctuations in pressure in turbopumps, and in designing new rockets. Most recently, the software helped to resolve a cracking issue in a valve in the Space Shuttle Endeavor's main propulsion system.

Due to the technology's benefits to space flight missions, the PKP-Module won Marshall Space Flight Center's "Software of the Year" award in 2009. That same year, the PKP-Module received a Space Act Award in the exceptional category by the Inventions and Contributions Board of NASA.

Product Outcome

With its roots in software development for NASA, ASRI has grown over the last 20 years to provide diverse engineering, programmatic and technical services not only to NASA, but also to the U.S. Department of Defense (DoD) as well as commercial customers. Today, there are more than 80 users of PC-SIGNAL who are enhancing mechanical design and development efforts across the country.

Made to use on a personal computer in a Microsoft Windows-based environment, the graphical user interface of PC-SIGNAL makes it easy to use, and the automated capabilities save time when analyzing large amounts of data. It can view signal properties at various domains through two-, three-, or four-dimensional displays. As Jong explains, "All of the mathematical routines, filtering, and enhancement processes to reduce unwanted noise and

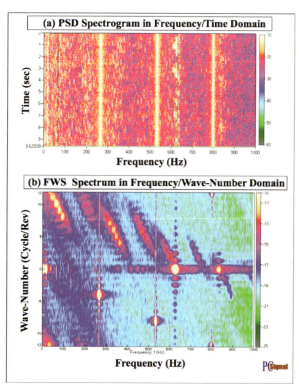

PC-SIGNAL created these dynamic displays of the signal properties in a space shuttle main engine.

other signal contamination are built into the programs. There is no additional programming necessary. Just point and click."

The software features a number of specialized techniques for engine and machinery diagnostic evaluation including vehicle bearing, gearbox, and drive train signal analysis. Applications include dynamic signal analysis, system health monitoring, flight data analysis, flow data analysis, and fatigue analysis and monitoring. The main markets are the aircraft and helicopter industries, rocket engine manufacturers, the transportation industry, and the nuclear power industry.

DoD employs PC-SIGNAL to calculate data about vehicle and helicopter components and materials including vibration specification development and fatigue analysis and monitoring. The U.S. Army Redstone Test Center uses PC-SIGNAL for vibration data processing and for developing specifications for its weapon system test programs.

The U.S. Air Force Research Laboratory has also applied PC-SIGNAL for vibration analysis in engine testing. Aside from government use, large aerospace firms such as Pratt & Whitney and Aerojet have purchased licenses from ASRI to use PC-SIGNAL for rocket and jet engine development, as well as for testing and analyzing the fluid flow in turbopumps.

While current applications of the software are largely focused on vibration analysis, opportunities exist for wider application of dynamic signal analysis in test facilities and laboratories. In the future, ASRI plans to provide a low-cost, portable, easy-to-use, real-time version of PC-SIGNAL for dynamic signal analysis, also based on the NASA-funded technology. ❖

PC-SIGNAL® is a registered trademark of AI Signal Research Inc.
Windows® is a registered trademark of Microsoft Corporation.

Modeling Tools Predict Flow in Fluid Dynamics

Originating Technology/NASA Contribution

Knowing what will happen before it happens is no easy task. That is why new spacecraft and technology are constantly being tested and refined—including the J-2X engine, which may power the upper stage of future NASA rockets. Data from tests like these help to ensure that the next generation of space explorers will travel safely into orbit.

Fueled by liquid oxygen and liquid hydrogen, the J-2X is a combination of the J-2 engine that propelled the Apollo-era rockets to the Moon, and the J-2S, a simplified version of the J-2 developed and tested toward the end of the Apollo Program.

The heart of this engine—the turbopump—is a complex piece of machinery that receives fuel from the storage tanks and then supplies it to the combustion chamber at the required flow, mixture, and pressure. Once the propellants mix and burn, they produce hot, pressurized exhaust gas that passes through a nozzle to accelerate the flow. This produces thrust that powers the rocket to move at approximately four times the speed of sound.

Because rocket engines like the J-2X operate under such extreme temperature and pressure, they present a unique challenge to designers who must test and simulate the technology. To model the physical and chemical phenomena that occur in these conditions, NASA engineers needed an advanced computer application with extremely complex and specialized capabilities.

Partnership

CRAFT Tech Inc., of Pipersville, Pennsylvania, specializes in computer programs for the simulation of complex fluid dynamics and combustion problems. In 2001, CRAFT Tech (an acronym for Combustion Research and Flow Technology) won the first of several **Small Business Innovation Research (SBIR)** contracts from Marshall Space Flight Center to develop software to simulate cryogenic fluid flows and related phenomena such as cavitation (bubbles that can lead to damage) in liquid rocket turbopumps.

According to Ashvin Hosangadi, principal scientist at CRAFT Tech, "There were a limited number of tools available to look at this problem at the time, but cavitation is very important to model, because it can cause major engine problems. For the first time, we were able to model the unsteady effects of cavitation in the low-pressure fuel pump of the space shuttle main engine (SSME)," says Hosangadi.

Through the SBIR program, CRAFT Tech enhanced its existing CRUNCH CFD (computational fluid dynamics) software and incorporated an advanced capability to predict the effects of cavitation in cryogenic fluids. This proved to be a significant improvement over the existing technology, which was limited to predicting turbopump performance in water.

By 2003, Stennis Space Center awarded CRAFT Tech additional SBIRs to extend the software's capabilities. To date, CRAFT Tech has established a solid foundation of state-of-the-art tools that have impacted the design and operation of rocket engine test facilities such as NASA's A-3 test stand for the J-2X and the B-2 test stand for the RS-68 five-engine cluster, as well as liquid propulsion for the J-2X and SSME.

"At Stennis, CRUNCH CFD provides a tool for designing reliable propulsion systems like the J-2X. By modeling rocket engine exhaust, or plumes, the software helps ensure those plumes will not cause damage to the facility, structure, or to the test article itself," says Dr. Daniel Allgood, aerospace engineer at Stennis.

Rockets of the future may be based on the Aries I-X, shown here at Kennedy Space Center in 2009. When developing the propulsion systems for future rockets, NASA uses a computer program from CRAFT Tech Inc.

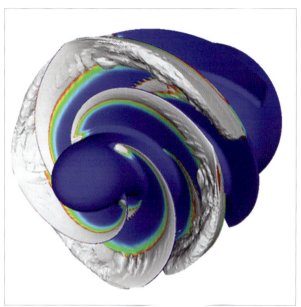

CRUNCH CFD software created this image of cavitation simulation of a cryogenic inducer, an important component of liquid rocket turbopumps.

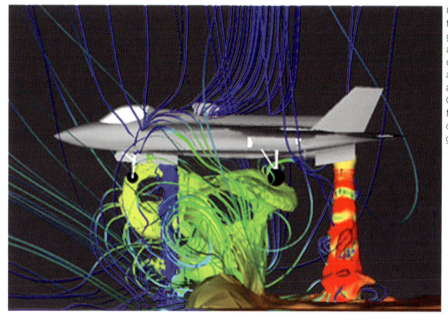

CRAFT Tech software created this visual representation of the ground effects on the exhaust from an aircraft. Specializing in fluid dynamics, combustion, and aeroacoustics modeling for aerospace and commercial applications, CRAFT Tech provides commercial tools, consulting support, system design and concept evaluation, and ground and flight test support.

Product Outcome

CRUNCH CFD can simulate the physical and chemical phenomena in various liquid propulsion components and systems including cavitating flows, supercritical flows at high pressures, gas and liquid mixtures at critical pressures, cryogenic and non-cryogenic liquid flows, and fluid combustion at high pressures.

The primary market for CRUNCH CFD is the commercial aerospace industry, but both government and industry clients use the software for analyzing existing systems as well as designing new ones. Northrop Grumman and Boeing use the program for cryogenic pump work, combustion problems, and modeling engine noise reduction.

CRUNCH CFD is most unique due to its advanced capability for predicting phenomena that happen with cryogenic liquids and in high-pressure systems. Hosangadi explains, "Liquid rocket systems work with cryogenic systems that operate at very high pressures. These can't be represented by simple equations used in other systems. Our specialty is to do simulations when representing the thermodynamics and physics becomes very difficult."

Like liquid rocket turbopumps, industrial commercial pumps also demand advanced modeling capabilities, including simulations under extreme pressure and temperatures. Using CRUNCH CFD, one of the largest pump companies in the United States has significantly improved the performance of its newer products. Today, the NASA-derived technology is being used with a range of large, high-energy pumps including boiler feed pumps in the utilities industries and oil pumps in the petrochemical industry.

A long list of potential applications exists, including a range of fluid flow devices such as valves and thermal heat exchangers. In particular, the liquefied natural gas industry is an ideal application for CRUNCH CFD. Supertankers that transport the cryogenic fluid are similar to liquid rocket cryogenic tanks, but on a bigger scale. One of the common problems is sloshing of the liquid, which can lead to system instability and safety concerns.

In addition to analysis, CRAFT Tech provides licensing of CRUNCH CFD for complex physics and thermochemistry applications, as well as performs fundamental research in CFD, grid adaptation, aeroacoustics, turbulence, combustion, and multiphase flow. The company, however, credits NASA for a large portion of its success. "The work on cavitation led to CRUNCH CFD being used in the commercial industry, thanks to the NASA SBIR program. All of our commercial revenue is very much a function of the work from SBIRs," says Hosangadi. ❖

CRUNCH CFD® is a registered trademark of CRAFT Tech Inc.

Verification Tools Secure Online Shopping, Banking

Originating Technology/NASA Contribution

Much is made of the engineering that enables the complex operations of a rover examining the surface of Mars—and rightly so. But even the most advanced robotics are useless if, when the rover rolls out onto the Martian soil, a software glitch causes a communications breakdown and leaves the robot frozen. Whether it is a Mars rover, a deep space probe, or a space shuttle, space operations require robust, practically fail-proof programming to ensure the safe and effective execution of mission-critical control systems.

Just as rovers are rigorously tested in simulated Martian conditions on Earth before actual mission launch, the software components must also be thoroughly analyzed to ensure the absence of bugs that might cause complications or critical failures. NASA's Robust Software Engineering (RSE) group, part of the Intelligent Systems Division at Ames Research Center, works to develop automated verification and validation technologies for aerospace and aviation software, making certain these programs are correctly written. In 1999, the group began developing a powerful verification tool, called Java Pathfinder (JPF), for programs written in the popular Java programming language.

"Java Pathfinder started as a model checker, which exhaustively goes through all of the possible behaviors of a program," says Corina Pasareanu, a Carnegie Mellon University and Ames RSE researcher. Since then, she says, JPF has evolved into a toolset including several technologies, one being a symbolic execution tool called Symbolic Pathfinder, which is the focus of Pasareanu's work. In 2005, JPF became one of the first NASA programs to be open-sourced, and JPF and Symbolic Pathfinder have garnered multiple awards, including ones from NASA, the Federal Laboratory Consortium, and IBM among others.

Partnership

JPF was the original tool for verifying native Java code. Fully testing Java code typically requires massive quantities of input to explore every possible program path—hence the Pathfinder name, also inspired by the Mars Pathfinder mission—and to verify the absence of bugs. These inputs are often generated manually, which requires significant time and effort. Symbolic Pathfinder is NASA's answer to that problem. The tool uses symbolic execution, in which inputs are specified as symbolic variables, able to take any value within a numeric range. The tool analyzes the code structure of the target program and generates symbolic constraints for the program's variables. These constraints are solved to automatically create test cases that reach all parts of the program code. Essentially, Symbolic Pathfinder has the capability to automatically execute a program on all possible inputs and in all possible ways to find defects and what causes them. The tool offers extreme thoroughness at less time, effort, and cost.

"We have applied this technology to several projects at NASA, generating test cases automatically and uncovering bugs in NASA software," says Pasareanu. Flight control software, scripts for commanding robots, and air traffic management software are a few of the NASA components that have benefited.

A significant deficiency with Symbolic Pathfinder, however, came to the fore when Fujitsu Laboratories of America Inc., based in Sunnyvale, California, began exploring the use of the open-source tool for testing Java programs for Web applications. Fujitsu's Software Validation Program, part of its Trusted Systems Innovation Group, works to develop verification and validation techniques to enhance public "trust" in information technology systems that play a major role in daily life, like wired and wireless networks, e-commerce, banking, and government systems, says Sreeranga Rajan, senior researcher for Fujitsu.

"The problem was that the NASA tool only handled integers, Booleans, real numbers, and recursive data structures as inputs. As soon as we tried to apply it to Web and enterprise applications, we found that the inputs are text strings," says Fujitsu researcher Indradeep Ghosh. Building on the Symbolic Pathfinder tool, Fujitsu created a new capability to handle text input variables as symbolic

Java Pathfinder has been used to analyze ground control software for the space shuttle.

inputs, thus allowing significantly broader verification of Java programs.

Product Outcome

If you do your banking or shopping online, you are most likely using Web sites enabled by Java applications. These applications (shopping carts, for example) can now be tested using Symbolic Pathfinder.

> *"This is the ideal case for government lab and industrial lab collaboration."*

"If you go to a banking application, you might have to input a check routing number, which is alphanumeric," says Rajan. Rajan explains that safety-critical classes of software, such as those involving financial transactions or for medical use, necessitate much deeper bug finding than typical testing techniques. Fujitsu's enhanced Symbolic Pathfinder tool, he says, "explores the program's behavior, analyzes the potential problem cases, and thereby helps cover a larger set of behaviors that typical testing programs cannot cover. That means we are able to unearth bugs that would not be found with existing technologies."

For commerce applications like shopping carts, Symbolic Pathfinder becomes a tool for ensuring security. With the new string input extension, Rajan says, Symbolic Pathfinder can help detect security holes that originate by string inputs.

Fujitsu plans to continue its work with JPF, scaling up its capabilities for verification of large-scale, complex Java programs. Rajan anticipates the company will use the toolset for in-house software development. The benefits of Fujitsu's work are also returning to NASA: The company has open-sourced its string input extension to allow the Ames JPF team and others to continue the tool's development.

"NASA was able to develop this complicated technology, something an industrial lab probably wouldn't have the time or resources to do," says Ghosh. "Once the technology matured, we could immediately see that, by improving it in a year or 2-year time frame, we could consider using it in an industrial setting. This is the ideal case for government lab and industrial lab collaboration."

Rajan notes that a significant amount of software development goes into security systems, and that the sharing of experience and capabilities between government and private industry can improve security for industry and for the Nation alike.

"NASA had developed in JPF something nobody else had developed for Java, and because of that we were able to get to the stage we are at now," he says.

This photo, taken during Mars Pathfinder testing, shows the Pathfinder lander opening its petals to expose the yellow Sojourner rover (on left petal). Just like robotics and other hardware, control software must be extensively tested to ensure proper execution of commands.

"Fujitsu Laboratories is glad to see this advancement in symbolic execution technology for strings transitioning to practicality," says Sanya Uehara, corporate senior vice president and director of Fujitsu Laboratories Ltd. "Such a milestone that has so far eluded verification researchers has been reached through collaboration between Fujitsu Laboratories in Japan and the United States, inspired by NASA's work on Symbolic Pathfinder." ❖

Toolsets Maintain Health of Complex Systems

Originating Technology/NASA Contribution

Monitoring the health of a machine can be just as tricky as monitoring the health of a human. Like in the human body, a variety of subsystems must work together for a machine to function properly—and a problem in one area can affect the well-being of another. For example, high blood pressure can weaken the arteries throughout the body, and weakened arteries can lead to a stroke or kidney damage. Just as a physician may prescribe medication, a special diet, or a certain exercise routine to maintain the health of a person, NASA employs a systems health management approach to ensure the successful operation of its rockets, crew vehicles, and other complex systems.

As a unique engineering discipline, systems health management prevents and minimizes problems in systems by using processes, techniques, and technology to design, analyze, build, and operate the systems. The goal is to understand a system, accurately assess its health (how well it is working), pinpoint any problems, and then support any maintenance or repair activities. Such an approach is critical to the success of NASA missions.

Partnership

To strengthen the systems health management approach used for NASA's large, complex, and interconnected systems, Ames Research Center awarded Qualtech Systems Inc. (QSI), of Wethersfield, Connecticut, several **Small Business Innovation Research (SBIR)** contracts. As a result, QSI adapted its Testability, Engineering, and Maintenance System (TEAMS) toolset and then employed the technology for the detection and isolation of electromechanical problems on the International Space Station (ISS) and again for the thrust vector control subsystem and ground hydraulics for the Ares I-X test flight in 2009.

First featured in *Spinoff* 2001, today the toolset includes TEAMS-Designer, a program that creates a

The TEAMS toolset was employed on the Ares I-X test vehicle, shown here at Kennedy Space Center's vehicle assembly building. TEAMS is also applicable to transportation systems, medical equipment, factory systems, telecommunications, and refinery systems.

model of the system and performs a large portion of the analysis; TEAMS-RT, a software embedded in the system for real-time monitoring and diagnosis; TEAMATE, software that communicates step-by-step guidance on how to troubleshoot a problem; and TEAMS-RDS, an enterprise-level server that enables maintenance and troubleshooting from a Web-based server, as well as a collection of field failure information for further improvement of troubleshooting strategy and to enhance future designs. Together, the toolset captures knowledge about how a system fails, how the failures are detected, and then uses that knowledge to guide engineers in troubleshooting and making real-time diagnoses.

Over the past decade, QSI has received significant funding through SBIR awards with NASA and the U.S. Department of Defense, and has advanced system diagnosis and health management, two of QSI's core capabilities. In 2002 and 2008, QSI received Space Act Awards from NASA, and in 2002, the *Aviation Week* award in recognition of the capabilities and advancements in TEAMS software. Now, Ames, Johnson Space Center, Kennedy Space Center, Glenn Research Center, Marshall Space Flight Center, and Jet Propulsion Laboratory all utilize TEAMS in support of NASA's Exploration Systems Mission Directorate.

Product Outcome

QSI has also adapted TEAMS for the business and technical requirements of large-scale commercial operations. After transitioning from a provider of consulting and engineering services to a provider of software products, QSI experienced revenue growth of more than 50 percent in less than 2 years, and also expanded its presence from the United States to Japan, Israel, and China.

Originally tailored for complex research applications at NASA, TEAMS can function in much the same way for simpler applications. Whether a system propels a spacecraft or an automobile, generates power, carries data, refines chemicals, performs medical functions, or produces semiconductor wafers, the conditions that cause the equipment to fail can be modeled and analyzed, then linked to test procedures, and finally generate a troubleshooting solution.

Dr. Somnath Deb, president of QSI, explains that a failure in one subsystem can propagate to other subsystems and generate false indications. "The subsystems are usually designed by different groups of people who don't know too much about each other's subsystems. TEAMS connects the dots and provides a comprehensive and accurate picture of the health of the entire system."

According to QSI data, TEAMS can substantially lower costs by decreasing warranty costs for manufacturers by 15 to 30 percent; diagnostic time across industries by 40 to 80 percent; calls to a call center by 30 to 60 percent; on-the-job training time by 30 to 60 percent; problems requiring field service by 30 to 50 percent; and repeat service calls by 70 to 90 percent.

In addition, QSI reports that one high tech company in the United States started utilizing TEAMS after comparing the time it took trained experts versus newly trained students to troubleshoot a mechanical issue. The student took 8 hours; the expert took 7.5 hours; using TEAMS, it took half an hour to troubleshoot the issue. According to QSI, field service engineers perform equally well when using TEAMS because it provides step-by-step instructions for troubleshooting.

A number of companies, including KLA-Tencor, Orbotech, Honeywell, General Motors (GM), Sikorsky, BAE, and Lockheed Martin are benefiting from using the NASA-derived technology for innovative new projects.

KLA-Tencor, a large semiconductor equipment manufacturer, is using the TEAMS tools for diagnosing and troubleshooting its suite of equipment at semiconductor fabrication sites around the world. The company expects a quick return on investment from its worldwide deployment of the TEAMS technology.

Orbotech, a supplier of inspection systems for printed circuit board production, deployed TEAMS in seven countries on three continents for its field service workforce. In this application, TEAMS averaged 75-percent accuracy on troubleshooting for new products without a maintenance history.

Under a subcontract for NASA, Honeywell selected TEAMS-RT for onboard vehicle health determination. Honeywell performed an evaluation of the competing technologies and selected TEAMS-RT for real-time diagnosis of the space flight vehicle Orion. Previously, Honeywell used TEAMS-RDS to monitor ISS data at Johnson, provide early detection and mitigation of problems, and supply continuous awareness of ISS health.

GM is also using TEAMS to evaluate how TEAMS-RDS could be used with its OnStar system, an in-vehicle security, communications, and diagnostics system. GM hopes to enable OnStar to read car engine codes and then tell the technician which part to repair. Deb explains, "This makes car maintenance more proactive because engine codes often show up before the car stalls or dies. Proactive maintenance would ensure the car drives well and the end user is not inconvenienced. Other companies are looking at TEAMS from this angle too."

According to Qualtech Systems Inc., TEAMS software can lower service costs by decreasing problems requiring field service by 30 to 50 percent. Shown here with accompanying test equipment called the enhanced turbine engine diagnostic system, TEAMS software is used on a laptop to model the T-700 helicopter engine in a test environment.

Fine-tuning its product with NASA has been extremely beneficial for QSI. "NASA wants the best, so it is open-minded and not only looks for the best, but also to bring out the best in you," says Deb. "NASA points out what to improve and how, and the SBIRs are a great vehicle for a small business to make improvements in a technology and still obtain ownership of it. That made a huge difference in making us who we are." ❖

TEAMS®, TEAMS-Designer®, TEAMS-RT®, TEAMATE®, and TEAMS-RDS® are registered trademarks of Qualtech Systems Inc. OnStar® is a registered trademark of OnStar LLC.

Framework Resources Multiply Computing Power

Originating Technology/NASA Contribution

For the last 25 years, the NASA Advanced Supercomputing (NAS) Division at Ames Research Center has provided extremely fast supercomputing resources, not only for NASA missions, but for scientific discoveries made outside of NASA as well. The computing environment at NAS includes four powerful high-performance computer systems: Pleiades, Columbia, Schirra, and RTJones. The collective capability of these supercomputers is immense, and in 2010, Pleiades was rated as the sixth most powerful computer in the world, based on a measure of the computer's rate of execution.

NASA was an early proponent of grid computing, which enables the sharing of computing power across geographic locations.

With loads of computing power—plus some to spare—NASA was an early proponent of something called grid computing, which enables the sharing of computing power, databases, and other online tools across geographic locations. The Information Power Grid was NASA's first distributed, heterogeneous computing infrastructure, linking computers at Ames, Langley Research Center, and Glenn Research Center into a connected grid.

Partnership

To demonstrate a virtual computer environment that links geographically dispersed computer systems over the Internet to help solve large computational problems, 3DGeo Development Inc., of Santa Clara, California, received **Small Business Innovation Research (SBIR)** funding from Ames in 2002.

3DGeo's work with Ames started with its Internet Seismic Processing (INSP) product. As a provider of advanced imaging software services for the oil and gas industry, 3DGeo developed INSP to provide an interface between clients and computer servers to manage advanced seismic imaging projects. Used to identify underground deposits of oil and gas, the seismic imaging process first sends sound waves to an area to create an echo that is recorded by an instrument called a geophone. The data from the geophone is then processed to create images of the subsurface.

At the time of the SBIRs, an existing key function of INSP was to facilitate the communication between members of an exploration team, providing the tools to easily share information, regardless of physical distance. To add to INSP's capability and to address the need for computational resources in excess of those locally available, 3DGeo developed a graphical user interface (GUI) for the detection of available remote resources for workflow design and execution, as well as for continuous monitoring of the resources being used. New features were built into INSP to extend its functionality for grid computing, and the technology was named "grid-enabled INSP," or G-INSP. The conversion of INSP to a grid-enabled system provided access to the resources needed to run advanced imaging applications whenever and wherever they were needed.

In 2008, 3DGeo merged with Fusion Geophysical LLC to become FusionGeo Inc., of The Woodlands, Texas.

Shown here is one row of cabinets that make up the 100-cabinet Pleiades supercomputer system at Ames Research Center. In 2010, Pleiades ranked as one of the fastest systems on the Top 500 list of the world's most powerful supercomputers.

Product Outcome

The NASA-derived G-INSP software is commercially available as a product called Accelerated Imaging and Modeling (AIM), which serves as a collaboration tool toward a fully optimized grid environment and provides a framework for integration at different stages of data processing. The software can create a virtual computer environment for any calculation-intensive and data-intensive application and is available as stand-alone software or as a component in other FusionGeo products.

Through its GUI, AIM offers functionality for searching for remote grid-resources; seeing and exploring file systems using grid file transfer protocol (FTP); visualizing data; dragging and dropping data sets between grid FTP file systems; creating and executing workflows on the grid; and overseeing workflows and checking intermediate results.

"Our product is an infrastructure for parallel computing, or high-performance computing, over the Internet. If you have a computationally intensive task, which requires thousands of central processing units (CPUs) of computing power, but you don't have that many CPUs—or don't have them locally—you can submit your job over the Internet and access computing power, storage, and data that are distributed at centers anywhere in the world," describes Dimitri Bevc, the chief technology officer at FusionGeo.

Because of the industry's requirements for very large processing and data storage capacities, AIM is particularly well-fit for seismic imaging for energy exploration and development. Based on data garnered through the seismic imaging process, the construction of accurate 3-D images of the subsurface is an extremely resource-intensive task. According to Bevc, it requires the handling of data on the order of 10 to 15 terabytes for a single marine 3-D survey. In addition, processing the data involves thousands of processors using computationally demanding algorithms. Only large-scale parallel computers can calculate these algorithms and deliver results within a useful time period.

Bevc explains that, without grid computing, a large seismic imaging project for the oil and gas exploration industry can take 1 to 2 years to complete. First, data tapes are transported physically between the data acquisition site, data banks, data processing sites, and quality control and interpretation sites, usually by the U.S. Postal Service, FedEx, or courier services. Once the data is interpreted, it is often reprocessed with a new set of parameters and then repeated. Ultimately, the process culminates with a decision to drill or not to drill a well.

AIM outfits oil and service companies with a grid-enabled virtual computer for larger 3-D seismic imaging projects, and provides computational resources on an as-needed, on-demand basis. "This can significantly shorten the time between receiving seismic data and making a drilling decision, and offers an ideal use of the grid to enhance the real-time value chain for end users in oil and gas exploration. Oil company experts, contractors, and service companies can collaborate between geographically disparate divisions and resources," says Bevc.

While AIM is currently used by oil companies and seismic service companies, the technology is also applicable to other computation- and data-demanding scientific applications such as geothermal energy exploration and for processes like carbon dioxide sequestration monitoring. ❖

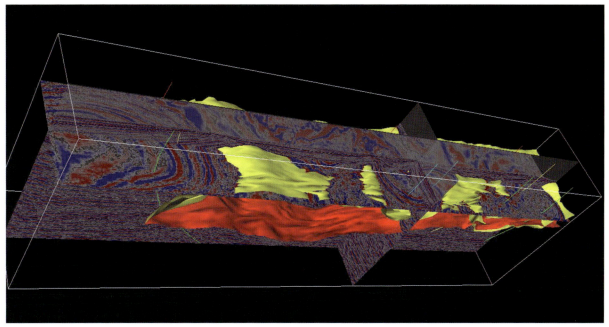

Major oil companies are utilizing FusionGeo Inc.'s software technology to construct accurate 3-D seismic images like the one shown here. Based on data captured through the seismic imaging process, the construction of these images of the subsurface is an extremely resource-intensive task.

Tools Automate Spacecraft Testing, Operation

Originating Technology/NASA Contribution

Using the Spitzer Space telescope, NASA scientists detected light from two Jupiter-sized extrasolar planets for the first time in 2005. Findings like these are enabled in part by the Science Mission Directorate at NASA, which conducts scientific research enabled by access to space—such as Earth science, planetary science, heliophysics (the study of the Sun and its effects on Earth and the solar system), and astrophysics (the study of the universe and Earth-like planets).

In 1988, NASA began a Small Explorer (SMEX) program to develop low-cost, high-performance spacecraft to advance the science of astrophysics and space physics. One of these spacecraft, the Fast Auroral Snapshot Explorer (FAST), probed the physical processes that produce auroras, the displays of light that appear in the upper atmosphere in the Earth's polar regions. Another spacecraft, the Transition Region and Coronal Explorer (TRACE), observed the connection between the Sun's magnetic fields and the heating of the Sun's corona. Additional spacecraft such as the Wide-Field Infrared Explorer (WIRE) studied the evolution of starburst galaxies and the search for protogalaxies.

To support these science-seeking missions, NASA began work on a software system to assist in developing and testing numerous types of spacecraft.

Partnership

As one of several commercial entities supporting the development of flight and ground software at Goddard Space Flight Center for NASA's SMEX missions, the Hammers Company Inc. (tHC Inc.), of Greenbelt, Maryland, created an all-in-one software for the development, testing, and on-mission operation of spacecraft.

"The notion was to develop a capability for the engineers to build and test the components before they were integrated on a spacecraft. In years past, we would build the spacecraft and integrate all the components. Then if there was a problem, we had to open up the equipment, remove components, and make repairs. The ability to test the system and components before integrating them has decreased the risk, costs, and maintenance schedule," describes Steve Hammers, a co-founder of tHC Inc.

Called the Integrated Test and Operations System (ITOS), the software successfully supported several SMEX spacecraft, and in 1999, the company exercised its option to make the ITOS software available commercially. Around the same time, tHC Inc. received **Small Business Innovation Research (SBIR)** funding from Goddard for a tool to facilitate the development, verification, and validation of flight software. Called VirtualSat, NASA has used the satellite simulation tool for various projects including the Earth Observer-1 mission.

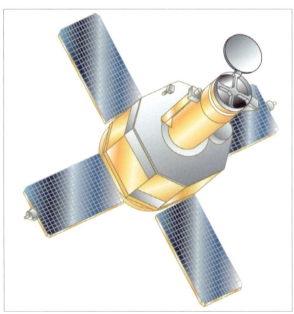

The Transition Region and Coronal Explorer (TRACE) spacecraft, shown here in an artist's rendering, launched in 1998 to explore the Sun and to define the upper solar atmosphere known as the Transition Region and Corona.

As a testament to the success of ITOS, NASA is currently using the software to support 15 orbiting satellites, and has 6 additional mission applications in development. Recently, ITOS passed Goddard Space Flight Center's orbital readiness review to manage three satellites in a multi-mission operations center. In addition, the company provided ITOS to the Lunar Reconnaissance Orbiter (LRO) mission experimenters to test instruments before launch.

Product Outcome

Together, the technology developed with NASA—ITOS and VirtualSat—has created and saved more than a dozen jobs for tHC Inc. Both ITOS and VirtualSat are currently being used by commercial entities in the aerospace industry.

A main advantage of the NASA-derived technology is its flexibility across different systems. "ITOS systems can be adapted and customized for a variety of requirements. The design is modular, so for every mission that comes up, we don't have to reinvent the wheel. It takes the system we have and modifies the components to meet new requirements," says Hammers.

ITOS and VirtualSat have been used to support the development of science instruments, spacecraft computer systems, command and data handling and guidance, and navigation and control software. By utilizing these tools early in the development cycle, the costs and risks associated with database conversions, training, validation, and maintenance can be reduced. Some of the entities using VirtualSat include Swales Aerospace (acquired by ATK), the U.S. Naval Research Laboratory, and Litton Industries (now a part of Northrop Grumman). Some of the entities using ITOS include the Canadian Space Agency, the University of California at Berkeley, and the Naval Research Laboratory.

As one of the company's first commercial clients, the Canadian Space Agency applies the software for multiple satellites. After 1 week of training, the agency was able

The Hammers Company Inc.'s (tHC Inc.) VirtualSat systems supported spacecraft flight software development and testing, as well as spacecraft environmental testing and checkout, at the Jet Propulsion Laboratory for NASA's Time History of Events and Macroscale Interactions during Substorms (THEMIS) mission.

Employed across the aerospace industry, the bridge in the lower right-hand corner of the image above was designed by tHC Inc.

to use ITOS independently. "We train users to make their own database in just a short period, and then they are off and running. A knowledge base is formed from instructions the operators make. Then ITOS will manage the commands, point the antenna, receive the telemetry, process the telemetry, send messages, and perform health and safety monitoring—all without human operators," says Hammers.

In support of NASA's Time History of Events and Macroscale Interactions during Substorms (THEMIS) mission, UC Berkeley manages a constellation of five satellites. There are usually no operators in the control room for THEMIS, because ITOS automatically sends commands to the antennas. The Universal Space Network, a provider of space operations, telemetry, tracking, and control services, also uses ITOS to control its antennas around the world.

As far as potential applications go, Hammers says ITOS could be used in factories and hospitals to monitor and control operations from a central location, as well as for controlling operations (like the opening and closing) of dams. "ITOS could potentially be used with any applications that have enormous amounts of data, typically engineering-related data, that needs to be received, processed, and monitored, and have the ability to communicate back to the devices," he says.

For both Hammers and his employees, NASA has been an inspiring force. "The synergy resulting from the partnership with NASA is more than just a business proposition. It creates excitement in our field that formulates new ideas, the development of leading-edge solutions, and draws some the best talent in the software field," says Hammers. "Our company is filled with very talented and excited employees, which is possible through our NASA relationship." ❖

GPS Software Packages Deliver Positioning Solutions

Originating Technology/NASA Contribution

To better understand and predict global climate, scientists look to the Earth's oceans. Natural forces like wind, storms, and heat affect ocean surface and sea level, and these changes can shed light on short- and long-term global climate patterns.

With the goal of tracking ocean currents and temperature over time, in 1979, NASA's Jet Propulsion Laboratory (JPL) started planning for TOPEX, a topography experiment to launch a satellite altimeter into space to measure the height of the world's oceans. Before scientists could launch TOPEX, however, they needed a way to obtain precise location information about where TOPEX was when it took measurements. To this end, JPL developed an innovative software program called the GPS (global positioning system)-Inferred Positioning System and Orbit Analysis Simulation Software, abbreviated as GIPSY-OASIS, to process and calculate data to determine a spacecraft's position.

In 1992, TOPEX was sent into orbit, and JPL achieved better-than-expected results. GIPSY-OASIS was able to pinpoint the location of TOPEX within just 2 centimeters. In addition, long-term results from TOPEX allowed scientists to observe El Niño, an oscillation of the ocean-atmosphere system characterized by unusually warm ocean temperatures, for the very first time.

Since the success of TOPEX, JPL has refined GIPSY-OASIS to become a sophisticated system that calculates accurate positioning information—not just for NASA, but for commercial entities as well. A companion software package called Real-Time GIPSY (RTG) was developed to provide positioning in certain time-critical applications.

Partnership

First featured in *Spinoff* 1999, the GIPSY and RTG software packages incorporate special GPS algorithms developed at JPL to deliver highly accurate positioning capabilities to a broad array of space, airborne, and terrestrial applications. In 2004, the precision GPS software was inducted into the Space Foundation's Space Technology Hall of Fame. Hundreds of commercial and non-commercial licenses for GIPSY and RTG have been released including more than 200 science and non-profit user licenses of GIPSY on a no-fee research basis.

One of the commercial organizations licensing GIPSY is Longmont, Colorado-based DigitalGlobe. In 2004, Doug Engelhardt, a principal systems engineer at DigitalGlobe, was looking for a new method for accurate orbit determination of the company's high-resolution imaging satellites. When he learned about the advanced capabilities of GIPSY, including its processing speed and accuracy, he decided to try it out.

Product Outcome

While TOPEX takes measurements of Earth's oceans, DigitalGlobe's satellites take pictures of Earth's surface. Like JPL, DigitalGlobe needs to know precisely where a satellite is when it gathers information. In order to place pictures accurately on a map, DigitalGlobe must know the location of the satellite when it shoots pictures of Earth, as well as the satellite direction when it took the picture. By combining this information, DigitalGlobe can assemble the pictures accurately to create high-resolution imagery of Earth.

Engelhardt explains that DigitalGlobe's satellites receive location information in space just like a hand-held GPS unit receives location information on Earth. The satellite then transmits the GPS location data as well as image data to one of the company's ground stations. It is sent back to DigitalGlobe's headquarters where GIPSY processes it. "As soon as we receive the data, the GIPSY process is kicked off. For every image that is processed, we need to know the precise location of the satellite, at the time it was taking the picture. After this is determined, the precise location data is linked with the images," he says.

Launched in 1992, TOPEX gathered data for determining seasonal cycles such as El Niño. To pinpoint where TOPEX was when it recorded information, NASA's Jet Propulsion Laboratory (JPL) developed GPS (global positioning system)-Inferred Positioning System and Orbit Analysis Simulation Software (GIPSY-OASIS).

By utilizing the licensed JPL technology, DigitalGlobe is able to produce imagery with highly precise latitude and longitude coordinates. This imagery is then made available to customers through an online platform and image library.

As one of two providers of high-resolution Earth imagery products and services, DigitalGlobe supplies imagery for a variety of uses within defense and intelligence, civil agencies, mapping and analysis, environmental monitoring, oil and gas exploration, infrastructure management, Internet portals, and navigation technology. As of March 2010, the company's content library had more than 1 billion square kilometers of Earth imagery, with 33 percent of it being less than 1 year old.

The company has a variety of city, state, and county government clients. Among other projects, these groups use the imagery for emergency response and infrastructure planning. "They keep track of the landscape in areas that are being developed with shopping centers, roads, and housing. They might look at the amount of trees and landscaping in an area, and then run computations on the amount that is paved, and how much runoff they need to account for in the drainage systems," describes Engelhardt.

Another major client of DigitalGlobe is Google, a provider of Internet search tools and services. Google uses the imagery for its Google Earth and Google Maps applications, providing a base layer of satellite imagery that Google can place overlay information on, such as roads, residences, and businesses. With Google Maps, users enter a single address or a starting and ending address, and then have the option to view the location or route with or without satellite pictures. If users choose the satellite view, they can see landmarks and other points of interest along the way. With Google Earth, users can enter an address or location and view, zoom-in, and fly through detailed satellite imagery of that specific location.

For a similar mapping platform, DigitalGlobe supplies imagery to Bing, a provider of Internet search tools and services from Microsoft. In addition, Nokia purchases DigitalGlobe imagery to use with its GPS service on select Nokia cell phones. Insurance companies also purchase DigitalGlobe's satellite imagery to get a distant look at damage resulting from natural disasters like hurricanes and floods, without having to visit the location. The news media are another frequent user of DigitalGlobe's satellite imagery to provide a bird's eye view of a newsworthy location.

When looking for a new method for accurate orbit determination of the satellites taking pictures like the one above of Washington, D.C., DigitalGlobe turned to JPL's GIPSY software. The software allows DigitalGlobe to make imagery with precise latitude and longitude coordinates.

Engelhardt says the NASA license has been invaluable to the company's success. "The capability provided by JPL has been huge. A vast amount of research and effort went into the software so it could be used by industry."

For the future, DigitalGlobe plans to focus on standardizing the imagery so it is easy for clients to use. "We have imagery and the customer has applications, but getting the formats to go between one another is a lot of work. As we build the commercial market, that is one of our big efforts. The government has been using satellite imagery for years, but for commercial customers, it is brand new." ❖

Google Earth™ and Google Maps™ are trademarks of Google Inc. Bing™ is a trademark of Microsoft Corporation.

Solid-State Recorders Enhance Scientific Data Collection

Originating Technology/NASA Contribution

On May 20, 1996, astronauts aboard the Space Shuttle Endeavor watched as a unique structure unfolded in space like a complex trick of origami. From the free-flying Spartan satellite the STS-77 crew had released from the shuttle's cargo hold, a massive circular antenna inflated into shape. About the size of a tennis court, the Inflatable Antenna Experiment (IAE) was the first space structure of its kind, laying the foundation for future work on inflatable satellites, telescopes, and even astronaut dwellings.

After the IAE's deployment, the structure was jettisoned and the Spartan satellite was recaptured by the shuttle. Key to the experiment's success was a new device onboard the satellite that recorded and stored the critical test data.

Partnership

Prior to the IAE, most NASA satellites and missions logged data using magnetic tape recorders. In terms of data gathering and storage, these missions were limited to how much these devices could record and how fast. The recorders' mechanical systems also made them vulnerable to failure; backup recorders took up valuable onboard real estate and added to cost and launch weight.

In 1983, National Reconnaissance Office pioneer Ray Anderson and son Scott founded SEAKR Engineering Inc., based in Centennial, Colorado. Ray Anderson had recognized that tape recorders represented a life-limiting spacecraft component, and SEAKR set about developing a replacement device that would lower cost while increasing reliability and capacity. The company turned to solid-state technology for a solution.

Solid-state, or flash memory, devices are crafted entirely from solid materials in which the only moving parts are electrons. A solid-state recorder (SSR) stores information in binary code (0s and 1s) in tiny transistors; transistors conducting current have a value of 1, while

The Inflatable Antenna Experiment, seen here fully deployed from its Spartan satellite, logged its data on an SBIR-derived solid-state recorder (SSR) developed by SEAKR Engineering Inc.

those that are not have a value of 0. Since there are no moving parts that could fail, and the information in the recorder can only be altered by voltage changes in the transistors, solid-state technology offers a more reliable, durable, and compact method of data storage. (Think of replacing a cassette player with an iPod nano.)

SEAKR first demonstrated its space-ready, SSR technology on a U.S. Department of Defense satellite before launching a device on the NASA-funded Orbital Sciences SeaStar satellite (now known as OrbView-2), which carried the Sea-viewing Wide Field-of-View Sensor (SeaWiFS) for imaging of the Earth's oceans. Under **Small Business Innovation Research (SBIR)** contracts with Goddard Space Flight Center, the company then built the SSR that replaced the original tape recorder on the Spartan satellite and gathered the test data for the IAE mission. SEAKR's NASA work helped the company advance its SSR technology.

"Demonstrating that you have a sound design that functions in space is a difficult hurdle to get over," says Scott Anderson, SEAKR's president. "Once we got over that hurdle, the fact that we were a small company with extremely low cost structures put us in a position to win opportunities as programs and customers switched from tape recorders to solid state."

SEAKR's SSRs opened up new capabilities for space missions that were previously hampered by tape recorder limitations.

"Suddenly, that bottleneck disappeared. Multiple, different instruments could record at the same time. They could run at considerably higher data rates," says Anderson. "The capacities of these solid-state systems considerably outperformed the tape recorders. They became mission-enabling technologies."

Product Outcome

Today, SEAKR is the world leader in solid-state recorder technology for spacecraft. The company has delivered more than 100 systems, more than 85 of which have launched onboard NASA, military, and commercial spacecraft, and none of which have experienced on-orbit failure. The company has also built upon its expertise with solid-state electronics to branch out into other critical flight technologies.

"If you think about what's in a solid-state recorder—a microprocessor, a lot of memory, software, packaging technologies—that's all the same technologies that's in a flight computer," Anderson says. "Building high performance computers for spacecraft is now our largest growing market segment." The company also offers bulk memory module products it says arose from technology developed for NASA programs like the Small Spacecraft Technology Initiative, the Mars Global Surveyor, and Mars Polar Lander.

SEAKR hardware has flown on many NASA missions, from satellites and planetary probes to the space shuttles and the Spitzer and Kepler space telescopes. The company's SSRs have recorded data for the Mars Global Surveyor and Mars Reconnaissance Orbiter, and SEAKR developed the High Rate Communication Outage Recorder (HCOR) for the International Space Station. The HCOR allows separate nodes within the station to route their science and communication data to a single, large memory buffer.

Outside of NASA, SEAKR has used its solid-state technology to build more than 1,100 memory boards for the U.S. Air Force C-17 Globemaster cargo aircraft. SEAKR SSRs are also part of the GeoEye-1 and WorldView-1 imaging satellites, which provide much of the high-resolution imagery for online mapping services.

"Whenever you see a picture on Google Earth, there's a good probability that the image came through one of our recorders," says Anderson.

SEAKR's technological accomplishments continue to push the limits of solid-state space electronics. Through the Internet Routing in Space project, the company collaborated with Cisco Systems and the Department of Defense and built a space-hardened Internet router that allows Web access and communications to occur directly through satellites rather than double-hopping to and from stations on Earth, a capability that may one day allow quick and highly reliable communications for soldiers in the field.

SEAKR is also working on another SBIR partnership with Goddard to build a high-performance flash-based memory module based on the VPX electronics standard, which allows for greater power and ruggedness. The results, Anderson says, will continue to advance the technology that demonstrated its usefulness so readily on the Spartan satellite nearly 15 years ago.

"This is going to be a great collaboration," says Anderson. "NASA is helping us fund a technology that is going to push the performance of these systems in the future." ❖

SEAKR's SSRs have become mission-enabling technologies, playing a vital role in many NASA missions and onboard multiple commercial satellites.

> *"NASA is helping us fund a technology that is going to push the performance of these systems in the future."*

iPod nano® is a registered trademark of Apple Inc.
Google Earth™ is a trademark of Google Inc.

Computer Models Simulate Fine Particle Dispersion

Originating Technology/NASA Contribution

Getting to the Moon is, to say the least, challenging. Being on the Moon, though, is no picnic either. In addition to the obvious life support and temperature control concerns, astronauts must contend with another obstacle: the Moon's surface. This surface is covered with sharp, abrasive dust, lunar soil, and rock, called regolith, which can pose a variety of problems for astronauts and their equipment.

When the Apollo astronauts touched down on the lunar surface, their visibility was severely limited due to dust being kicked up, a phenomenon similar to the "brownout" conditions helicopter pilots experience when flying low in deserts—but coupled with the concern that this abrasive ejecta could damage the lunar module, their only chance of returning home safely. Once on the surface, the astronauts had to contend with the dangers of inhaling or ingesting this fine dust and the possibilities that that their equipment could be damaged or contaminated. These fears were not without warrant; a camera failed when lunar dust entered a drive mechanism during the Apollo 15 mission.

Any return trip to the Moon's surface, whether it involves humans or robots, will necessitate new methods for predicting the role that lunar regolith will play in the mission, whether it is through surface contamination, excavation, mining, or handling of the dust and lunar rock. Understanding more about the chemical and physical properties of lunar regolith, like how particles interact with each other and with equipment surfaces, or the role of static electricity build-up on dust particles in the lunar environment, is imperative to the development of technologies for removing and preventing dust accumulation and successfully handling lunar regolith.

At NASA's Kennedy Space Center, Dr. Carlos Calle, founder and director of the Center's Electrostatics and Surface Physics Laboratory, is currently working on these problems: the electrostatic phenomena of granular and bulk materials as they apply to planetary surfaces.

Through a NASA Seed Fund partnership, Calle was able to partner with industry to enhance commercial modeling software, with mutual benefit to both NASA's mission needs as well as those of its industry partner.

Partnership

To accomplish some of its technology initiatives, NASA is relying on some innovative methods, such as the Centennial Challenges, a series of prize competitions aimed at spurring inventors and design teams to help NASA design the technologies of the future, with chances to win large cash prizes in the process. The Lunar Lander challenge, for example, is a multiphase competition that challenges teams to build a vertical takeoff vehicle/vertical landing rocket capable of simulating the same characteristics needed for movement between the Moon's surface and its orbit. Another competition in the Centennial Challenges program is the Lunar Regolith Excavation Challenge, which calls upon teams to build a robot capable of excavating simulated regolith, given weight and power restrictions. Both of these challenges seek to leverage the know-how and creative spirit of amateur and student inventors to benefit NASA's future mission needs.

Using EDEM software to simulate lunar dust behavior, NASA investigated how to efficiently clean dust away from equipment on the Moon or Mars—by using the influence of an electrical field to repel the particles. The arrows show dust moving away from a screen that needed cleaning.

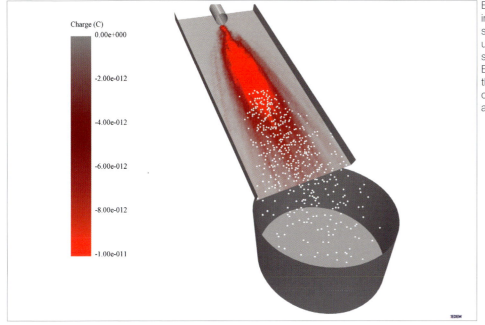

EDEM simulation investigates how materials such as lunar soils will build up electrical charge while sliding down a chute. With EDEM, NASA can analyze the effects of different chute materials on charge accumulation.

An additional innovative method NASA has recently developed for leveraging expertise to meet its needs is the Seed Fund program, which provides funding to address barriers and initiate cost-shared, joint-development partnerships. One such Seed Fund partnership was with DEM Solutions Inc., of Lebanon, New Hampshire. DEM Solutions is a leader in particle dynamics simulation software. Its EDEM software is a computer-aided engineering (CAE) tool that uses state-of-the-art discrete element method (DEM) technology for the simulation and analysis of bulk materials flow in a wide range of particle handling and manufacturing operations. EDEM allows users to import computer-aided drawing models of particles, to obtain accurate representations of their shapes, and to assign physical properties so that the models replicate the behavior of real bulk solid materials.

This software, the most advanced DEM technology commercially available, showed great promise for the study of lunar regolith, but before it could be applied to the NASA work, additional modeling and validation of electrostatic and surface energy adhesion forces, dynamic friction, and particle shape were required. Through the Seed Fund, NASA and DEM Solutions worked together to enhance the existing software to provide more accurate modeling of lunar regolith.

The DEM Solutions Seed Fund proposal provided NASA with both the needed expertise in particle dynamics simulation and a customized, easy-to-use CAE tool for qualifying dust/coating interaction by virtually modeling the behavior of the dust in contact with coatings of electrodynamics screens. EDEM simulations are now routinely used in the development of NASA surface systems dust mitigation technologies designed for dust removal and cleaning of optical and thermal radiator surfaces, connectors, and seals. EDEM virtual prototyping capabilities also streamline the design process for development of systems for In Situ Resource Utilization (ISRU) excavation, traction, material flow, and the electrostatic separation of ilmenite, a mineral found in lunar regolith. "We appreciated this opportunity to expand our software platform to meet NASA needs," says John Favier, DEM Solutions CEO. "We are always pleased to develop advanced simulation solutions for our customers, but it is especially gratifying to contribute to the NASA mission of space exploration, technology development, and resource utilization."

The enhancements made during this partnership have been of interest to DEM Solutions' customers, as well, and the software developments made during the NASA partnership are also finding their way into diverse industry settings.

Product Outcome

These new EDEM capabilities can be applied in many industries unrelated to space exploration and have been adopted by several prominent U.S. companies, including John Deere, Pfizer, and Procter & Gamble.

For John Deere, the NASA and DEM Solutions technology is of interest, as the analysis and modeling of fine, angular, and cohesive particles will enable the company to improve its handling of similar materials on Earth, like sandy or cohesive soils.

In the pharmaceutical realm, the software is able to help manufacturers deal with the fine particles necessary for drug manufacturing, where highly active ingredients, often in fine powder form, need to be contained, maneuvered, and mixed. Accurate modeling of particulate behavior makes this process safer and more efficient. With the newfound, NASA-enabled features, the already one-of-its-kind EDEM software is sure to make future terrestrial and space-related modeling of fine particles a much more accurate and reliable endeavor. ❖

EDEM® is a registered trademark of DEM Solutions Inc.

Industrial Productivity

The engineering lessons from NASA missions and research help to build things other than space vehicles. The technologies featured in this section:

- Lighten Components
- Reveal Elements in the Short Wave Infrared
- Correct Optical Distortions
- Advance Optics Manufacturing
- Integrate Optical Functions
- Automate Processes, Catalyst Testing
- Record Unique Videos of Space Missions
- Purify Semiconductor Materials
- Control Motion of Complex Robotics

Composite Sandwich Technologies Lighten Components

Originating Technology/NASA Contribution

At Glenn Research Center's Ballistic Impact Facility, engineers study new materials and the ways that they react to sudden, brute force. Using high-speed cameras, various sizes of gas-powered guns, and a variety of other tools, these engineers learn about impacts, ways to build stronger, lighter materials, and how to avoid catastrophic events caused by high-pressure collisions. It was this laboratory that investigated the accident caused by the external tank foam striking the leading edge of the Space Shuttle Columbia's wing, which led to that orbiter destructing during atmospheric reentry, causing the loss of the spacecraft and crew. The lab's findings were integral to the safety changes initiated before the shuttle fleet was again operational.

Other ballistic impact studies include examining various new materials, composites, and methods for creating jet engine fan casings. With the rotor fan blades spinning at very high speeds, if there is a failure—perhaps caused by a foreign object being sucked into the fan duct—and a piece or all of a blade breaks free, it shoots out like a bullet. This can cause catastrophic damage to the engines or passengers. To avoid this, NASA works with industry to develop strong, lightweight materials that can replace the traditional steel and aluminum jet engine fan casing.

Partnership

Leveraging its private resources with several **Small Business Innovation Research (SBIR)** contracts with both NASA and the U.S. Department of Defense, WebCore Technologies LLC, of Miamisburg, Ohio, developed a fiber-reinforced foam sandwich panel it calls TYCOR that can be used for a wide variety of industrial and consumer applications. Testing at the Ballistic Impact Facility demonstrated that the technology was able to exhibit excellent damage localization and stiffness during impact.

According to Mike Sheppard, WebCore program manager, the SBIRs with NASA "allowed WebCore to evaluate the TYCOR product family in a variety of applications, specifically fan cases and launch vehicle components." Through the NASA testing, the company has been able to use "'real world' loads and criteria to size our products effectively; arriving at optimized sandwich panel solutions. The input from NASA relative to requirements and the feedback that the NASA team provided relative to testing plans and threshold criteria have been invaluable."

Product Outcome

The TYCOR fiber-reinforced composite is strong, lightweight material ideal for structural applications (*Spinoff* 2004). The patented and trademarked material has found use in many demanding applications, including marine, ground transportation, mobile shelters, bridges, and most notably, wind turbines.

In marine applications, where it is prized for its corrosion-resistance, strength, and low weight, TYCOR can be used for hulls, decks, interiors, and hard tops on small

Developed in part with SBIR funding from NASA, TYCOR is an alternative to balsa and foam cores, helping blade designers and fabricators reduce total cost and weight.

leisure craft, as well as ships and larger boats. Because of the material's ability to withstand the jostling, vibrations, and structural demands of long term travel, it can be found in semi truck floors, side panels, roofs, doors, and ramps, as well as truck bodies and railcar floors. TYCOR has also been used to build mobile specialty shelters; the material is ideal for mobile units because of its light weight and strength. WebCore has also used TYCOR in multiple bridge and infrastructure projects where specific strength, stiffness, and performance metrics were required, and corrosion and high maintenance and installation costs were factors to be avoided.

One of the most recent applications for TYCOR is in the field of wind turbines, where, according to Sheppard, "TYCOR offers a compelling value proposition which stresses not only competitive pricing but also weight savings, resin savings and overall manufacturing cost reduction when compared to other sandwich core materials."

The lightweight composite allows builders of turbines to make larger, more efficient turbines. And with its low price, TYCOR allows users of wind energy to recoup their costs more quickly. "TYCOR's largest market, today, is wind energy. WebCore supplies our core material to wind blade manufacturers around the Nation and around the world," says Sheppard.

This year, the technology was selected by NASA's Advanced Composites Technologies (ACT) Project of the Exploration Technology Development Program (ETDP) as one of two final candidates for the acreage panel structure of a novel "beamless" core inter-tank structure design for heavy lift vehicles. ETDP develops new technologies that will enable NASA to reduce the cost and risk involved with future human space flight missions. Under the ACT Project, seven different technologies were being tested and examined for their applicability in future space missions.

"This was a major milestone for WebCore, certainly, but it also represented an opportunity for NASA and the broader aerospace community to move outside historical

One of the prime applications for TYCOR is in wind turbine blade manufacturing, where its design flexibility allows blade designers and manufacturers to tailor modifications for individual blade and turbine specifications.

constraints such as those associated with autoclave processing. This shift has the potential to open the door to larger, more weight-efficient structures that may result in lower cost, payload mass, and cost structures, overall."

During the ACT Project, WebCore worked with Collier Research Corporation (*Spinoff* 2009) to develop a TYCOR module for Collier's NASA-derived HyperSizer program, which helps designers optimize complex composite structure designs. Licensed from NASA's Langley Research Center, the software helps lighten and strengthen structures by allowing users to experiment with new, composite materials. The first-ever NASA software license, HyperSizer, has been increasingly reused by NASA as the Agency has been innovating new spacecraft designs.

According to Sheppard, "Collier worked with WebCore to incorporate a TYCOR module in the HyperSizer tool which allowed for independent sizing, analysis, and confirmation of WebCore's value in launch vehicles and other structures." ❖

TYCOR® is a registered trademark of WebCore Technologies LLC.

Cameras Reveal Elements in the Short Wave Infrared

Originating Technology/NASA Contribution

In late 2009, a rocket traveling twice as fast as a speeding bullet crashed into the Moon as part of NASA's Lunar Crater Observation and Sensing Satellite (LCROSS) mission. The resulting impact loosened a mixture of particles, dust, and debris that was analyzed by a host of instruments to confirm the presence of water on the Moon. Two of these instruments were indium gallium arsenide (InGaAs) short wave infrared (SWIR) cameras, used to image the wavelengths between visible and thermal on the electromagnetic spectrum.

Employing InGaAs technology commercialized by Greg Olsen (who also happens to be the third private citizen to go into space and orbit Earth), the cameras were ideal for NASA's application because SWIR light tends to be absorbed by moisture, creating contrast in the image between the darker, moister elements and the brighter, drier elements.

Partnership

Before being acquired by Goodrich Corporation in 2005, Sensors Unlimited Inc., based out of Princeton, New Jersey, worked under a variety of **Small Business Innovation Research (SBIR)** contracts with NASA's Jet Propulsion Laboratory, Marshall Space Flight Center, Kennedy Space Center, Goddard Space Flight Center, Ames Research Center, Stennis Space Center, and Langley Research Center. Since 1991, the 25-plus NASA SBIR contracts helped to spur state-of-the-art research and development to advance and refine the InGaAs imaging technology that was ultimately used on the LCROSS mission. In addition to NASA, other supporters of SWIR technology include the U.S. Army Night Vision Electronic Sensors Directorate and the Defense Advanced Research Projects Agency. Today, the technology also has dozens of commercial applications.

"NASA has been very influential in getting some of the first devices to market, specifically the active pixel

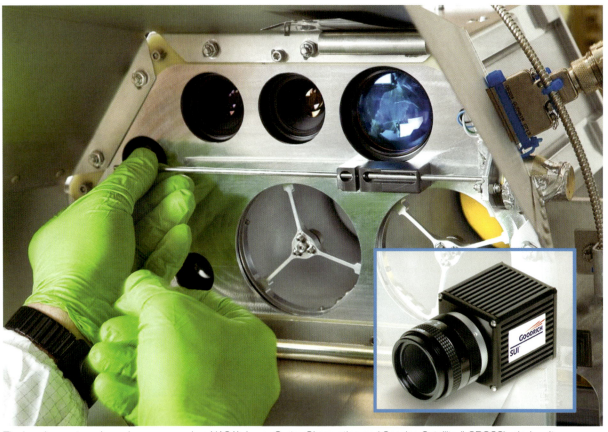

The imaging system shown here was used on NASA's Lunar Crater Observation and Sensing Satellite (LCROSS) mission. It incorporated several cameras to capture the impact of a rocket crashing into the Moon. Two of these cameras were Goodrich Corporation short wave infrared (SWIR) cameras, like the one shown in the lower right-hand corner.

sensor incorporated in most all InGaAs detector arrays," says Robert Struthers, director of sales and marketing for Goodrich ISR Systems in Princeton, New Jersey (formerly Sensors Unlimited Inc.).

In 1998, the company's work with NASA resulted in the first InGaAs camera to image both SWIR and visible light from the same image sensor. "This is a valuable device for compact imaging spectrometers (an instrument that measures light). If you don't have room for multiple cameras, you can have one camera cover both bands," says Struthers.

Product Outcome

Struthers explains the fundamental element of the imaging technology is the photodiode. Basically, a photodiode converts a quantity of light (photon) to a quantity of electricity (electron) to a grayscale image. "In our spectral band, we are looking at reflective light, just like visible

These images of San Francisco Bay show the difference between images produced by a visible light camera and a SWIR camera. The visible light image (left) shows mostly fog; the SWIR image (right) reveals detail through the fog. SWIR technology detects reflected light at wavelengths that the human eye cannot see, in bands on the electromagnetic spectrum between the visible and thermal.

light cameras do. Once you convert light to electricity, then you run that into an amplifier to increase the signal or into an image processor to create the image. We do all this in a camera head so all the customer has to do is plug the camera into a monitor," says Struthers.

While other cameras can operate in the SWIR range using sensors constructed from mercury cadmium telluride or indium antimonide, these cameras must be mechanically cooled, often to extremely low temperatures. In contrast, the InGaAs cameras can operate at room temperature. In addition to its lack of a cooling system requirement, what makes the InGaAs camera so unique is its small size, low weight, high sensitivity, high resolution, day to night imaging capabilities, and low power consumption.

Today, all of Goodrich ISR Systems' products out of its Princeton location are based on the InGaAs material, the mainstay technology for imaging in the SWIR range.

Even though there are fewer than a dozen main products, there are hundreds of variations of each.

By providing a variety of customized products to fit clients' needs, Goodrich ISR Systems has expanded the set of applications for SWIR instruments, such as detecting moisture in pulp and paper and agriculture, and for sorting recycled plastic products. Applications also abound in sectors such as the military, security and surveillance, machine vision, medical, spectroscopy, semiconductor inspection, instrumentation, thermography, and telecommunications.

In the agriculture, textile processing, and forest-product industries, moisture is a key indicator for process control and quality. Because water is opaque to SWIR illumination, detecting its presence or absence can be useful in gauging crop health and product ripeness or dryness. Using the SWIR cameras, elements with higher water composition appear darker than drier elements.

For the telecommunications industry, Goodrich ISR Systems' products are incorporated into photodiode arrays that are used as performance monitors for fiber optic networks. The light transported across the fiber optic cable in telecommunication networks is detected within the SWIR or near-infrared spectrums. The technology is integrated into a larger optical power monitor or optical channel monitor to ensure high-quality optical signal and data integrity, and can be found in many of the world's optical networks.

In 2009, the cameras were widely used for inspection systems to determine defects in solar panels. For solar cell manufacturers, solar cell inspection by machine vision with InGaAs SWIR permits detection of defects, hidden cracks, or saw marks inside or on the opposite side of the wafer. Beyond physical defects, SWIR solar cell inspection can find problems that will hurt cell or system power output.

"This wavelength can image right through silicone, so when you want to inspect anything made out of silicone like image sensors, microprocessors, or solar panels, the SWIR is ideal," says Struthers.

As for future SWIR products, Goodrich ISR Systems is currently working on a very small SWIR InGaAs camera, as well as the first high-speed, high-resolution SWIR InGaAs linescan camera. "We're expanding our technology to make even higher resolution image sensors, similar to high-definition TV format sensors with InGaAs," says Struthers.

Regardless of where Goodrich ISR Systems goes, it will never forget where it came from. "Working with NASA has helped the advancement of our device performance, including some very valuable steps along the way," says Struthers. "The SBIR program is ideal for seeding small technology companies like we used to be." ❖

Deformable Mirrors Correct Optical Distortions

Originating Technology/NASA Contribution

In March of 2009, the Kepler spacecraft was launched to explore the structure and diversity of planetary systems outside of our own solar system, with a special emphasis on the detection of Earth-sized planets. Once Kepler fulfills its mission, the SIM Lite spacecraft will follow in its path to measure the masses of the planets discovered by Kepler and to study additional planets. Together, these missions will pave the way for the Terrestrial Planet Finder (TPF), an instrument designed to have imaging power 100 times greater than the Hubble Space Telescope, to ultimately provide the first photographs of new planetary systems.

Currently under study at NASA, the TPF will combine the high sensitivity of space telescopes with revolutionary imaging technologies consisting primarily of adaptive optics. To this end, a very small yet extremely important piece of technology is taking shape: the deformable mirror (DM).

As one of the main elements in adaptive optical systems, DMs adjust their shape or position to correct for aberration (an optical phenomenon that leads to blurring or distortion). Originally proposed in the 1950s for improving astronomical imaging, and developed by the U.S. Air Force in the 1970s, adaptive optical systems consist of three main elements: a DM, a sensor that measures the aberration hundreds of times per second, and a control system that receives the measurements and then tells the DM how and where to move in order to correct for the distortion. Today, major Earth-based telescopes such as the Keck Observatory in Hawaii and the European Southern Observatory in Chile rely on adaptive optics to remove distortion caused by light passing through Earth's atmosphere. For space applications, adaptive optics correct for distortions in the telescope due to temperature, launch impact, and absence of gravity.

Partnership

In 1999, Boston Micromachines Corporation, a newly formed company in Cambridge, Massachusetts, received a Phase I **Small Business Innovation Research (SBIR)** contract with the Jet Propulsion Laboratory (JPL) for space-based adaptive optical technology. The Phase I led to a Phase II SBIR with JPL and resulted in a microelectromechanical systems (MEMS) DM called the Kilo-DM.

"Space-based telescopes need a small mirror that can correct for aberrations. MEMS are a smart choice, because they are compact, low-power, and lightweight," says Paul Bierden, president and co-founder of Boston Micromachines. MEMS technology integrates tiny actuators and electrical interconnects on one silicon chip. Electronics are connected to the actuators to control their position.

From 1999 to 2009, the company gained manufacturing expertise based on additional SBIRs with JPL. The batch manufacturing approach developed for the small and compact devices was much less expensive than competing technologies. Because the MEMS DMs are low power, the company could make the electronics smaller, allowing for easier integration, which also contributed to less costly manufacturing. "The SBIR program has accelerated the development of this technology by years and has allowed us to get devices into people's hands much faster," says Bierden.

Product Outcome

Boston Micromachines now features a full product line of MEMS DMs, including the Kilo-DM, the Multi-DM, and the Mini-DM. The difference among the products is the number of actuators, or elements that change the

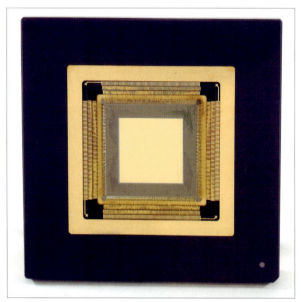

Boston Micromachines Corporation gained manufacturing expertise through SBIRs with the Jet Propulsion Laboratory for its microelectromechanical systems (MEMS) Kilo-deformable mirror (DM). As part of adaptive optical systems, DMs adjust their shape or position to produce clearer images.

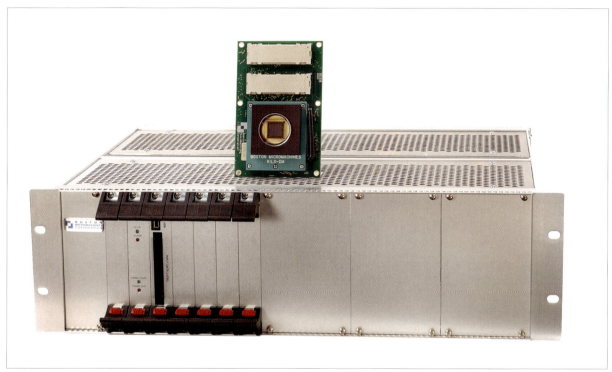

The Kilo-DM and high-speed electronics driver are ideal for demanding applications in astronomy and laser communication. Currently, the European Southern Observatory, NASA's Terrestrial Planet Finder (under development), and various defense agencies are using the Kilo-DM.

reflective surface. Ranging from 32 to 1,020, the more actuators a DM has, the greater the image correction.

Observatories around the world are using the Kilo-DM for astronomical imaging. At Mauna Kea, Hawaii, it was integrated in the Subaru Telescope, operated by the National Astronomical Observatory of Japan, to achieve better images of stars and distant galaxies. It also provides clearer views for observatories in Arizona, France, and the United Kingdom. Most recently, the technology was customized for the Gemini Planet Imager, the next-generation adaptive optics instrument being built for deployment at the Gemini Observatory in Chile. Operated by a partnership of seven countries, the goal of the Gemini Planet Imager is to look at extrasolar planets orbiting nearby stars.

In addition to widespread astronomical applications, the technology has been adapted to improve the quality of laser communication systems. Used primarily by defense organizations, laser communication systems (wireless communication through the atmosphere using a laser beam) are limited in their speed and accuracy due to turbulence and other interference in the air between the sender and receiver. Scientists are finding that by using adaptive optical technology, distortion can be corrected, resulting in a higher data rate over a longer range. "NASA funded the development of the Kilo-DM and then Boston Micromachines took that to a Defense Advanced Research Projects Agency (DARPA) program," Bierden explains. "They needed the technology to be modified, so we built a product for DARPA based on what we developed with NASA."

Another Earth-based application for Boston Micromachines' technology is microscopy. In combination with fast-scanning lens technology, a team at Rensselaer Polytechnic Institute incorporated the Mini-DM into an adaptive scanning optical microscope (ASOM) to get a clearer look at biological samples. Thorlabs Inc. has licensed the ASOM and made it commercially available for use in the fields of biotechnology, medicine, industrial quality assurance, and automated medical diagnostics.

As for future applications, vision science researchers are testing the MEMS DMs for their ability to eliminate blurriness caused by eye tissue and to get a better look at the retina. This is encouraging for the early detection and diagnosis of eye conditions such as glaucoma, diabetic retinopathy, and macular degeneration. The MEMS DM also looks promising for noninvasive deep tissue in vivo imaging as well as endoscopic imaging.

The company is on the verge of commercializing the newest technology developed through an SBIR contract with the Jet Propulsion Laboratory: a MEMS tip-tilt piston (TTP) DM. This device is a tiny segmented mirror in which 331 independent segments can pivot on their axes to correct for aberrations. By leveraging technology currently in development for the Gemini Planet Imager, the company is now focused on increasing the number of elements in their TTP MEMS DM to 1,027 for space-based imaging, which could lead to both power lasers and optical lithography applications.

"NASA SBIRs have not only allowed Boston Micromachines to do fundamental research, move our company and technology forward, and provide solutions for NASA, but also to provide solutions for other people in the imaging community," Bierden says. ❖

Stitching Techniques Advance Optics Manufacturing

Originating Technology/NASA Contribution

The amount of detail a telescope can see is directly related to the size of its mirrors. To look deep into space at galaxies over 13 billion light years away, NASA requires telescopes with very large mirrors. Scheduled to launch in 2014, the James Webb Space Telescope (JWST) will have one 6.5-meter primary mirror as well as one 0.5-meter secondary mirror.

While most optical surfaces are spherical, the primary and secondary mirrors in the JWST are aspheres, or non-spherical. Such optics are far more difficult to manufacture, but the quality of images seen through an asphere is far superior to those seen through traditional spherical lenses.

To make a high-precision optics element, the surface must be precisely measured and then polished to meet those measurements. The use of a precision tool called an interferometer (which measures the surface) is important, because an optical surface can only be polished as precisely as it can be measured. For the highest quality optics, the surface is measured and polished to the nanometer level, or to 1/100,000th the size of a human hair. When it comes to measuring aspheres, often custom interferometers have to be built for each specific application.

Because NASA programs depend on the fabrication and testing of large, high-quality aspheric optics for applications like the JWST, it sought an improved method for measuring large aspheres.

Partnership

QED Technologies, of Rochester, New York, had developed a device for measuring spherical optical surfaces up to and exceeding approximately 0.2 meters. Called a subaperture stitching interferometer (SSI), the technology used a stitching technique that measured a series of sections on an optical surface, compiled data from each section, and then stitched it together to obtain a complete measurement. This enabled more accurate and full-aperture measurements of large optics.

Intrigued by the technology, Goddard Space Flight Center awarded two **Small Business Innovation Research (SBIR)** contracts to QED to upgrade and enhance the efficiency, capability, and flexibility of the stitching technology for aspheres.

As a result of working with NASA, as well as follow-on work with the U.S. Army Research Laboratory, QED developed the SSI-A, which was capable of applying the stitching technique to larger aspheres. The company also developed a breakthrough machine tool called the aspheric stitching interferometer (ASI), the first-ever commercial solution that enabled the measurement of high-departure aspheres without the use of dedicated null optics (additional optics that are expensive, difficult to make, and often inaccurate).

Dr. Andrew Kulawiec, metrology product line manager at QED, explains the benefits of the technology developed in part with NASA. "It reduces the cost and

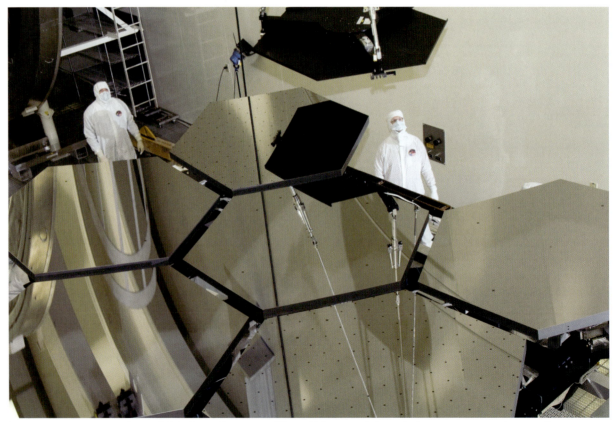

The primary mirror for the James Webb Space Telescope will have 18 hexagonal-shaped mirror segments, each measuring 1.3 meters in diameter. Here, the mirror segments are being prepped to move into the X-ray and Cryogenic Facility at NASA's Marshall Space Flight Center.

lead time for producing aspheres compared to traditional methods. It's beneficial, because you can buy this instrument to measure a specific asphere, and when you are done with it, you can measure the next asphere. For our customers, the companies who are making aspheres, it provides a lot of flexibility to be able to go after new business," says Kulawiec.

In 2008, the SSI-A earned QED the "R&D 100" award, granted by *R&D Magazine* for being one of the 100 most technologically significant products introduced into the market that year. Then in 2010, QED was named a finalist for the "Prism Award for Photonics Innovation" for its ASI.

Product Outcome

High-precision optical elements are used not only in telescopes, but also in microscopes, cameras, medical scopes, and binoculars. While a typical microscope objective may have 6 or 8 elements, a professional camera may have 10 to 20 optics. To achieve a sharper image or to look at a larger area, even more optics may be added to a device. Another option is to incorporate aspheric optics into the device.

"In conventional optics, all of the surfaces have a spherical shape, but if you change some of those surfaces to have an aspherical shape, you can reduce the number of elements that are required to achieve the same image quality, and potentially reduce costs as well," says Kulawiec.

Until recently, the manufacturing of aspheres was time consuming and expensive, with each asphere requiring a custom interferometer setup. With QED's advances, one device can be used to measure many different optics, so a variety of industries have started to incorporate aspheres into their imaging devices.

"The ASI has allowed aspheres to become more prevalent in the optics industry by making them easier to manufacture and to measure," says Kulawiec. "NASA enabled the optics industry to make and measure the types and parts that it needed."

Compared to traditional methods, QED Technologies' Aspheric Stitching Interferometer decreases the cost and lead time for producing aspheres (non-spherical optics). By using more aspheres in an optical system like a microscope, designers can decrease the total number of optical elements, weight, and size.

The use of aspheres results in higher quality images, fewer optical elements, less weight, and reduced system size. As a testament to the popularity of aspheres, QED's ASI has remained in a sold-out state since it was introduced to the market in 2009. QED has sold systems in Asia, Europe, and North America for research and development, military and defense, security and surveillance, and digital imaging.

One of the leading suppliers of optics and optical components, Edmund Optics (EO), uses the SSI-A for the correction of high-precision prisms and aspherical surfaces. According to the company, QED's technology has allowed for rapid measurement of aspherical surfaces as well as allowed EO to meet and exceed its customers' application needs.

An additional application for the technology is in photolithography, or the process used to print patterns on the computer chips used in electronics like cell phones and laptops. QED's systems assist the industry in manufacturing the high-precision optics used in this process.

In 2010, the ASI was implemented at the Leibniz Institute for Surface Modification in Leipzig, Germany, for precision fabrication and measurement of aspheres. According to the institute, the ASI is an important addition, because it can develop surface figuring and finishing steps that produce surfaces with accuracies in the nanometer range.

As for the future of the technology, QED plans to scale it up for even larger optics. "The ASI is a significant innovation and new capability for manufacturing aspheres, but we don't believe it ends there. We believe further development is possible to measure even more complex shapes. It's a technology we can work with and continue to develop for quite some time." ❖

SSI-A® and ASI® are registered trademarks of QED Technologies.

Compact, Robust Chips Integrate Optical Functions

Originating Technology/NASA Contribution

People often think of NASA's research as pushing the boundaries of our universe—sending people, robots, and spacecraft into the dark reaches of the sky. But as NASA studies the universe, one of the most important planets it explores is Earth. Through a series of Earth-observing satellites, high-altitude research aircraft, and ground-based validation methods, NASA unveils many of the secrets of Earth's climate, assessing everything from changes in ocean temperature and phytoplankton health, to the melting of the polar ice caps, to the levels of particulate matter in the atmosphere.

To assist the Agency in its ongoing atmospheric studies, engineers at NASA's Langley Research Center developed the High Spectral Resolution Lidar (HSRL) to measure cloud and aerosol properties. These measurements are useful in the study of climate and air quality, while providing validation for Earth-observing satellites. Flown on Langley's King Air B200 aircraft, HSRL is designed to take vertical measurements of air particles, examining the air's content at different heights to give researchers a full depth of knowledge about a particular air space, a feat unable to be accomplished by satellites, for example, which, while able to provide a wide perspective, are only capable of examining the top, visible layer of a given area.

HSRL has been used to validate the measurements made by NASA's Cloud-Aerosol Lidar and Infrared Pathfinder Satellite Observation (CALIPSO) satellite-based lidar instrument. A joint venture between NASA and CNES, the French space agency, CALIPSO is investigating the ways that clouds and airborne particles affect the planet's weather, climate, and air quality. HSRL has also been used to monitor air quality and particulate matter over the Caribbean, Mexico, Alaska, and Canada, and was used to study wildfires threatening South Carolina's Myrtle Beach. By taking vertical measurements of the smoke, the HSRL was able to demonstrate how smoke plumes behave—how the smoke rises and then drifts through the atmosphere—which helps researchers gain a better understanding of how wildfires affect air quality and climate systems.

Partnership

Located in Bozeman, Montana, AdvR Inc. provides engineered optical crystals, lasers, electro-optics products and integrated fiber-based systems to government, academic, and university researchers. Since 1999, AdvR has been assisting NASA in developing technologies for a variety of missions through the Agency's **Small Business Innovation Research (SBIR)** and **Small Business Technology Transfer (STTR)** programs. In addition to working with Langley and NASA's Goddard Space Flight Center, the 15-person company has also worked with the National Science Foundation, the U.S. Army, U.S. Navy, as well as with numerous academic and industry partners.

Langley engineers partnered with AdvR through the SBIR program to develop new, compact, lightweight electro-optic components for remote sensing systems. Through this SBIR work, AdvR created a technology that will aid HSRL in making precise measurements, while saving space and weight. The technology: an optical waveguide circuit. (A waveguide is a physical device that steers waves, such as sound waves or radar waves, in a specific direction. An optical waveguide guides electromagnetic waves in the optical spectrum. In this case, it is a fiber-coupled engineered crystal that guides the light.)

The circuit integrates several key functions into a single, compact, robust, rugged device while still producing the precise measurements necessary for NASA's use. As such, AdvR's optical waveguide circuit has been integrated into the new laser transmitter system being designed for HSRL and that is currently being tested in the laboratory, with anticipated incorporation into a test flight this year on the B200 research aircraft.

Product Outcome

AdvR's circuit chip combines two optical functions—conversion of infrared light to the visible spectrum, and spectral formatting of the light so it can be locked to a precise wavelength for the high resolution and accuracy measurements required for HSRL. These functions are combined into a device about the size of a matchbox, which reduces the volume of the standard equipment by about a factor of 10. The size reduction is due in part to the optics being fiber coupled, as well as to the integration of several key functions in a single component.

Similar devices that rely on free-space beams of light are limited in how the light can be directed and are susceptible to vibrations, a problem on devices designed for aircraft. The AdvR integrated system utilizes all fiber-coupled components, which can be easily routed within a compact housing, allowing for the reduction in size and weight while also enabling highly accurate data collection. According to Shirley McNeil, senior laser systems engineer at AdvR, "There's nothing else on the market that combines these critical electro-optic functions into a

Through the SBIR program at Langley Research Center, AdvR Inc. developed components for lidar-based remote sensing instruments.

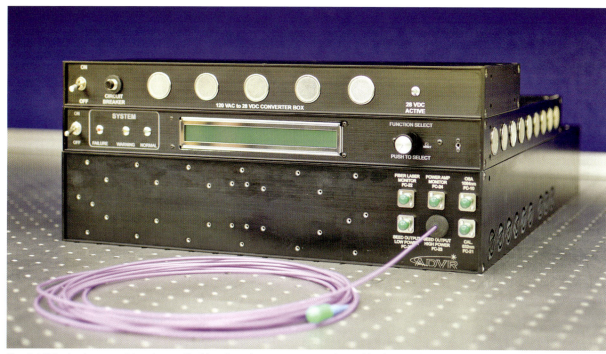

The AdvR technology combines two optical functions in one compact, robust chip that precisely locks the laser to the required wavelength. Precise locking of the laser and filter wavelengths provides greater accuracy.

> *AdvR foresees additional uses for its NASA-derived circuit chip anywhere that compact, low-cost, stabilized single-frequency lasers are needed.*

single, compact, low loss component. We are excited to see this technology in actual use on HSRL flight missions."

These qualities make the device useful for Langley's HSRL, but also for use on other NASA missions, such as some of the lidar remote sensing projects out of Goddard Space Flight Center and Langley's Differential Absorption Lidar program, which is currently using five wavelengths and could potentially use this device to convert some of the wavelengths, therefore requiring fewer sources. And while the primary customer for this technology will be NASA, AdvR foresees additional uses for its NASA-derived circuit chip in the fields of academic and industrial research—anywhere that compact, low-cost, stabilized single-frequency lasers are needed.

One such adopter of the technology is a German company, Berlin-based PicoQuant GmbH. PicoQuant uses one of the components AdvR initially developed under the SBIR effort—the frequency converter—and commercialized a 530-nanometer pulsed module for a variety of research applications, such as time-correlated single photon counting, a technique to record low-level light signals with picosecond time resolution; fluorescence correlation spectroscopy; microscopy; and other bioscience and medical applications.

According to McNeil, "One such biomedical use of this pulsed laser source is to study the blood coagulation process at a molecular level." A group of researchers led by Dr. Sandy Ross at the National Institutes of Health-funded Center of Biomedical Research Excellence in Missoula, Montana, tags the blood cells with fluorescent dyes, which are essentially probes for the interaction of interest. When pulsed light of the proper excitation wavelength illuminates the probes, the probes glow with lifetimes of a few nanoseconds, and they are able to characterize their time-dependent behavior of the processes that initiate the biological process of interest. A new pulsed source at 530nm allows Ross to more fully characterize this process, and by characterizing and understanding the blood coagulation process, they hope to provide information that will help enhance or slow down the coagulation process for specific medical applications.

Most of these applications focus on the device's ability to convert infrared lasers to visible wavelengths, a characteristic that could lead to this technology's adoption for displays, including large billboards and televisions. Laser-based displays and TVs use less energy than LCD or plasma screens while providing much improved contrast ratio and resolution, but the difficulty is in generating a laser in the visible green spectrum. Three primary colors (red, blue, and green) are required to generate all the colors in the visible spectrum. Red lasers, similar to those used in the bar code scanners at the grocery store, are readily available. Direct generation of blue is now also available, but direct generation of the green spectrum has not been developed. The most efficient method currently available is high-efficient frequency conversions of the infrared to the green. Significant progress is being made on laser-based TV and displays, and AdvR's technology could be utilized for this market. ❖

Fuel Cell Stations Automate Processes, Catalyst Testing

Originating Technology/NASA Contribution

Since the early days of space flight, fuel cells have provided an important source of power for extended missions to space. Aboard the space shuttle, three fuel cell power plants generate all of the electrical power from launch through landing.

After the Space Program refined them, fuel cells achieved widespread recognition by academia, industry, and the government as a means of producing clean energy. Compared to conventional combustion-based technologies, fuel cells produce much smaller quantities of greenhouse gasses and none of the air pollutants that create smog or cause health problems. When pure hydrogen is used as a fuel, fuel cells emit only heat and water as byproducts. In fact, the fuel cells on the space shuttle produce potable water that is pumped into storage tanks for the crew to use in orbit.

At the center of NASA's fuel cell research and development, Glenn Research Center performs work that could lead to new flight capabilities, electric power for long-term human exploration beyond Earth orbit, more efficient cars and trucks, and a cleaner environment. Glenn continues to look for ways to improve fuel cells—including the equipment used to test fuel cell technology.

Partnership

In 1991, Glenn awarded a Phase II **Small Business Innovation Research (SBIR)** award to Lynntech Inc., of College Station, Texas. Working with Glenn, Lynntech addressed one of the major limitations of fuel cell testing equipment by developing equipment and techniques to fully automate the fuel cell testing process. Five years later, the company was issued a patent on the breakthrough technology and began to offer the test equipment to the commercial world.

By 1999, the equipment was being sold internationally, and as demand grew, Lynntech Industries was formed in 2001 to concentrate on the manufacturing, sales, marketing, and servicing of the equipment, which was separate from the technology development efforts of the company. Not only did Lynntech's customer base grow, but the company's product line grew in the fuel cell and catalyst support industry—and extended to all fuel cell types, all power levels, and all stages in the development and testing cycle, from concept through production testing.

By 2004, Lynntech made the decision to transfer its successful fuel cell technology, patents, and personnel to a spinoff company called Fideris. Shortly after Fideris introduced highly modular, plug-and-play, rack-mounted components into its product line, ownership of the Fideris technology and brand was transferred to TesSol Inc., of Battle Ground, Washington. TesSol, which stands for Testing Solutions, is a provider of the Fideris brand testing

Space Shuttle Endeavour rests on a launch pad at Kennedy Space Center before the STS-118 mission. Fuels cells onboard Endeavor produce all of the electrical power needed for its mission.

equipment to fuel cell researchers and developers, catalyst companies, and research centers around the world.

Product Outcome

Building on the invention made under NASA's SBIR program, the Fideris product line continues to offer computer-controlled, fully automated fuel cell test equipment with ultra-low impedance electronic loads, real-time evaporation humidifiers, a graphical user interface, and all-digital communication with remote monitoring and control. Whether for a fuel cell, catalyst, battery, or process control; in a laboratory, or in a manufacturing environment; the technology provides automated operation and control of the process, device, or material being tested. Applied to fuel cell testing, for example, the test equipment provides automated fuel, oxidant, and temperature control while measuring and monitoring various parameters such as pressure, temperature, and power output.

Because of its versatility, the technology has also expanded into a variety of other applications, and a large percentage of the equipment is used for non-fuel cell work such as catalyst testing, environmental chamber atmosphere control, sensor testing, chemical reactors, stack and exhaust emission control technologies, gas blending and humidification, temperature control, and the automation of other physical processes.

Companies are using the Fideris equipment with NASA-derived technology in laboratories to evaluate the performance of products, as part of testing services, for quality control in production environments, and for general process control and automation. The Fideris equipment is found at universities, national laboratories, government entities, and small and large businesses around the world.

For the testing of catalysts, sensors, and stack emissions treatment technologies, the system evaluates the performance of the catalyst under various conditions as well as for its lifetime. In one application, a supplier to the

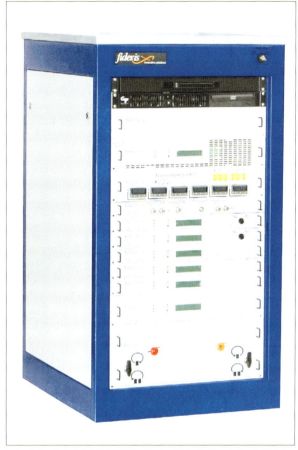

Fideris testing modules satisfy the testing needs of researchers and manufacturers in the fields of energy, fuel cells, catalysts, and sensors. The flexibility of the equipment lends itself to field reconfiguration, hardware, redundancy, and endurance testing.

auto industry uses the equipment to evaluate the underbody catalyst used for exhaust treatment in automobiles. In another, a commercial test facility uses the technology to perform catalyst evaluation as an independent testing service. The technology is also being used by suppliers to test NOx and O_2 sensors, as well as to test ambient gas detectors and personal monitoring detectors.

> *Thanks in part to the NASA SBIR program, the technology has saved money for the fuel cell industry.*

Now a full 15 years after the original concepts were patented, the technology continues to be one of the most critical aspects of fuel cell testing. The widest application is in gas blending and humidification in laboratories, including the feeding of a humidified gas stream across the product or device to evaluate corrosion, to regulate how a sample is dried, to modify the composition of the sample, or to grow surface layers. Fideris' equipment can also maintain the temperature of environmental chambers, evaporation processes, reactors, and dryers. In addition, it can be used for thermal management for performing automated temperature profiles of baths, and for blending liquids or gasses to perform automatic concentration sweeps for a reaction or process.

Thanks in part to the developments made possible through the NASA SBIR program, Dr. Craig Andrews, chief technology officer of TesSol, says the technology has saved money for the fuel cell industry.

"The technology opened the door to automated testing of fuel cells. As the fuel cell industry matured, it became more important to perform endurance testing as well as maintain a testing program where a specific test is repeated in the exact manner for direct comparison," says Andrews. "Our equipment has reduced the manpower required for testing and turned away from the old 'one-person, one test stand' to suites of test stands that can perform tests for months or years without intervention." ❖

Fideris™ is a trademark of TesSol Inc.

Onboard Systems Record Unique Videos of Space Missions

Originating Technology/NASA Contribution

It was one of the few times that a crash landing would be deemed a success. On October 9, 2009, nine sensor instruments—including five cameras—onboard the Lunar Crater Observation and Sensing Satellite (LCROSS) watched closely as the Moon-bound spacecraft released the spent upper stage of its Centaur launch vehicle at the lunar surface. The instrument-bearing shepherding spacecraft beamed back video of the Centaur's impact and then descended through the resulting plume, gathering data on the composition of the ejected material until it too impacted within the lunar crater Cabeus. The mission yielded a wealth of information confirming what scientists had hoped for: the presence of water on the Moon.

A specially designed avionics unit controlled, routed, and transmitted back to Earth the precious data gathered from all of LCROSS' onboard instruments. The crucial control unit was the outcome of a collaboration between NASA's Ames Research Center and a unique company whose products have benefited from this and other NASA partnerships.

Partnership

In 1999, a company called BlastOff! Corporation was formed in Pasadena, California, with the intent of landing the first commercial robotic lander, equipped with a host of onboard video and imaging systems, on the Moon. The company folded in 2001, but a group of BlastOff! employees went on to form Ecliptic Enterprises Corporation, also of Pasadena, to take advantage of the expertise they developed in creating ruggedized video systems for use in space.

Onboard video systems for rockets or spacecraft provide stunning footage of launches and space activities—valuable material for educating and inspiring interest in space exploration. But another significant benefit is the essential information these video feeds provide to engineers on the ground. While casual viewers get to experience a virtual ride into space, watching the Earth fall away under a rocket's flames, engineers gain important situational awareness, allowing them to monitor and evaluate a rocket launch or the activity of a complicated mechanical device on a spacecraft.

The need for comprehensive situational awareness became readily apparent in the aftermath of the Columbia disaster. The Space Shuttle Columbia broke up while reentering Earth's atmosphere during its 2003 mission, killing its seven crewmembers. Investigators concluded the shuttle's destruction was caused by hot gasses entering through a hole in the thermal protection of the vehicle's left wing; the hole was caused by the impact of a chunk of foam insulation that broke away from the shuttle's external fuel tank during launch.

Seeking ways to improve situational awareness for future shuttle launches, NASA examined the use of multiple onboard cameras. On the STS-114 Return to Flight mission, Ecliptic's external tank camera captured the breakaway of a large piece of insulating foam, an incident which again grounded the shuttle fleet for nearly a year until the problem could be resolved. Of the multiple cameras trained on the shuttle during the launch, only Ecliptic's onboard camera provided the precise time the foam broke away from the tank, says company CEO and former Jet Propulsion Laboratory engineer Rex Ridenoure.

"The value of onboard video for situational awareness really got a boost in the aftermath of the Columbia tragedy," he says. Now the shuttle features multiple camera systems mounted on the external tank and on the solid rocket boosters.

Ecliptic RocketCam systems were incorporated into multiple other NASA missions—including the Delta II rockets for the 2003 twin Mars Exploration Rover missions and on the Demonstration of Autonomous Rendezvous Technology (DART) spacecraft in 2005—before the company signed a Memorandum of Understanding with Ames in 2007 to collaborate on

The science payload panel (left) for LCROSS (in the artist rendering above) underwent final testing at Ames Research Center in late 2007. The Data Handling Unit built by Ecliptic Enterprises Corporation is the gold box near the center.

Smaller, more capable RocketCam Digital Video Systems such as this will be used onboard future NASA, commercial, and defense rockets and spacecraft.

projects for onboard imaging systems and related technologies. As part of this collaboration, Ecliptic opened a small office in the Ames-based NASA Research Park.

Ecliptic became involved in the LCROSS mission when Ames principal investigator Anthony Colaprete realized the company's digital video controller—the avionics unit of its RocketCam Digital Video System (DVS) technology—could serve as the core technology for the spacecraft's Data Handling Unit, providing cost-effective control capabilities for its sensors.

"Up until this time, our video controllers were only controlling video cameras and other imaging sensors," says Ridenoure. "LCROSS wanted us to control several other sensors we had never seen, with lots of switching between them, lots of data formatting." The demands of the LCROSS mission required Ecliptic to develop a controller capable of handling higher data rates, more frequent sensor switching, and more data formatting complexities than its previous systems.

"LCROSS helped us push the capabilities of these digital video controllers and set us up to start tackling high-speed and high-definition video," Ridenoure says.

Ecliptic was able to advance its technology even further following the successful LCROSS mission. The company collaborated with NASA's Dryden Flight Research Center to develop a high-speed video system for monitoring the parachute deployments for the Constellation Abort Flight Test program, designed to test a launch abort system for the Orion crew capsule. Ridenoure says Ecliptic's work with Dryden developed the company's high-speed video capabilities and primed its technology for high-definition applications, since high-speed and high-definition share similar data rates.

Product Outcome

To date, Ecliptic's analog and digital RocketCam systems have been employed on more than 80 rocket launches and spacecraft missions for customers including NASA, the U.S. Department of Defense, and multiple aerospace companies. The company's technology has captured unique perspectives of an array of rocket launches, including Delta IIs, IIIs, and IVs; Atlas IIs, IIIs, and Vs; Titan IVs; and Minotaur Is and IVs. Ecliptic video systems also allowed the world to share the experience of Scaled Composite's SpaceShipOne aircraft making the first-ever privately funded human space flight. The company's systems are part of launches roughly every 4 to 6 weeks—with 8 to 12 in a typical year.

Ecliptic does not manufacture cameras, but rather takes off-the-shelf sensors, ruggedizes them to withstand the extreme conditions of launches and space operations, and houses them in protective pods and other enclosures that are affixed to the rocket or spacecraft. The company's key technology is its digital video controller, which is why Ridenoure is quick to note that Ecliptic is not a "camera company."

"Ninety percent of what we do is avionics, sensor-handling, and data switching, like what we did for LCROSS," he says.

Thanks to the company's work on the LCROSS mission and the Abort Flight Test program, Ecliptic has gained not only desirable technical capabilities but also caché among commercial spacecraft developers.

"From a commercialization angle, it's largely because our systems were baselined and approved for various challenging NASA missions that commercial satellite programs have confidence in our systems for their spacecraft," Ridenoure says. Ecliptic systems with "lots of heritage with the one we had on LCROSS" are now on geosynchronous commercial satellites. The company has also generated a preliminary design for a high-definition video system based on the experience it gained from its Ames and Dryden work and anticipates its first sale in this category within the next year.

As NASA and commercial space partners develop new vehicles for traveling into low Earth orbit and beyond, Ecliptic expects to provide the video footage that will keep engineers apprised and the public in awe. Ecliptic systems are set to be incorporated on the Orbital Sciences Cygnus vehicle and Taurus II rocket, designed to ferry cargo to the International Space Station (ISS) as part of the Commercial Orbital Transportation Services program. This year, the company received its largest contract ever to supply the United Space Alliance with RocketCam DVS technology for solid rocket boosters on future NASA launch vehicles.

Ecliptic will also enable a major educational and public outreach project when it launches onboard the two Gravity Recovery and Interior Laboratory (GRAIL) spacecraft, set to map the gravitational field of the Moon in late 2011. MoonKAM (Moon Knowledge Acquired by Middle School Students) will be composed of Ecliptic video systems on each spacecraft and is sponsored by Sally Ride Science, led by the former astronaut who also initiated the ISS EarthKAM. Students will be able to schedule video recordings and retrieve the clips from NASA's datastream for educational activities, hopefully inspiring a new generation of space explorers. ❖

RocketCam™ is a trademark of Ecliptic Enterprises Corporation.

Space Research Results Purify Semiconductor Materials

Originating Technology/NASA Contribution

While President Obama's news that NASA would encourage private companies to develop vehicles to take NASA into space may have come as a surprise to some, NASA has always encouraged private companies to invest in space. More than two decades ago, NASA established Commercial Space Centers across the United States to encourage industry to use space as a place to conduct research and to apply NASA technology to Earth applications. Although the centers are no longer funded by NASA, the advances enabled by that previous funding are still impacting us all today.

For example, the Space Vacuum Epitaxy Center (SVEC) at the University of Houston, one of the 17 Commercial Space Centers, had a mission to create advanced thin film semiconductor materials and devices through the use of vacuum growth technologies—both on Earth and in space. Making thin film materials in a vacuum (low-pressure environment) is advantageous over making them in normal atmospheric pressures, because contamination floating in the air is lessened in a vacuum.

To grow semiconductor crystals, researchers at SVEC utilized epitaxy—the process of depositing a thin layer of material on top of another thin layer of material. On Earth, this process took place in a vacuum chamber in a clean room lab. For space, the researchers developed something called the Wake Shield Facility (WSF), a 12-foot-diameter disk-shaped platform designed to grow thin film materials using the low-pressure environment in the wake of the space shuttle. Behind an orbiting space shuttle, the vacuum levels are thousands of times better than in the best vacuum chambers on Earth.

Throughout the 1990s, the WSF flew on three space shuttle missions as a series of proof-of-concept missions. These experiments are a lasting testament to the success of the shuttle program and resulted in the development of the first thin film materials made in the vacuum of space,

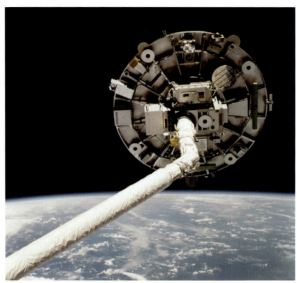

Before being released by Space Shuttle Endeavour during the STS-69 mission in 1995, the Wake Shield Facility (WSF) lingers above Earth.

helping to pave the way for better thin film development on Earth.

Partnership

In partnership with Johnson Space Center, researchers at the SVEC dedicated a significant amount of time to identifying which and exactly how the crystals should be grown on the WSF. After years in the lab preparing for the launch of the facility, researchers advanced an existing epitaxy technique called molecular beam epitaxy (MBE). In 1997, researchers from the SVEC formed a company called Applied Optoelectronics Incorporated (AOI), of Sugar Land, Texas, to fabricate devices using the advanced techniques and knowledge.

"Research in the lab at the Commercial Space Center, in preparation for on-orbit research, allowed some advances that wouldn't have been discovered otherwise," says Stefan Murry, vice president of global sales and marketing at AOI. "By doing research for the WSF, we advanced the state of the art in MBE. It's still the foundation technology that AOI uses to make all of its laser products."

Product Outcome

Today, AOI manufactures a crystal material used to make laser diodes and photo diodes. Laser diodes are placed on one end of an optical fiber to convert electrical energy into light energy while photo diodes are placed on the other end of an optical fiber to convert light back into electrical energy.

Murry explains, "Anything that involves sending large amounts of data over an optical fiber typically requires a laser diode and a photo diode. Basically, the laser diode transmits the information and the photo diode receives the information. We also make electronic modules that power and condition the signals that are driving the diodes so they can send and receive data more efficiently."

More than a decade after its start-up, AOI has become one of the largest developers and manufacturers of advanced optical devices—including laser diodes, photo diodes, related modules and circuitry, and other equipment for fiber optic networks including cable television, wireless, telecommunications, data communications, and fiber-to-the-home applications—selling more than 1 million devices each year.

"I don't think anyone planned to develop a great laser technology, but as a consequence of the Commercial Space Center, it happened. We found an application for something we were doing in the lab and were able to make it available commercially. As a result of developing technologies that were of interest to NASA, the commercial application presented itself," says Murry.

Now, AOI's products are an important component of broadband networks around the world, with a customer base in over 24 countries composed of equipment manufacturers like Cisco Systems, Motorola, Alcatel-Lucent, and Huawei. Particularly attractive for outdoor applications, one of the main advantages of AOI products is their ability to withstand high temperatures.

The company is also delivering some of its products to NASA for research and development. The Jet Propulsion Laboratory is using AOI's products for research on the deep space network, a large and sensitive scientific telecommunications system for tracking spacecraft.

AOI also supplies a substantial number of devices for fiber-to-the-home applications. "In the beginning, fiber optics was used for large amounts of traffic between telephone companies. As time went on, the ultra-high data carrying capacity became more important and moved out to consumers. Now optical fiber is used for home applications," says Murry.

While Murry believes residential fiber optics networks will continue to keep the company busy, he says AOI is also focusing on increasing the data-carrying capacity of laser diodes. Currently, the state of the art for laser diodes is about 10 billion bits per second—or 10 billion flashes of light per second. AOI is looking to improve that capacity by 4 or 10 times. ❖

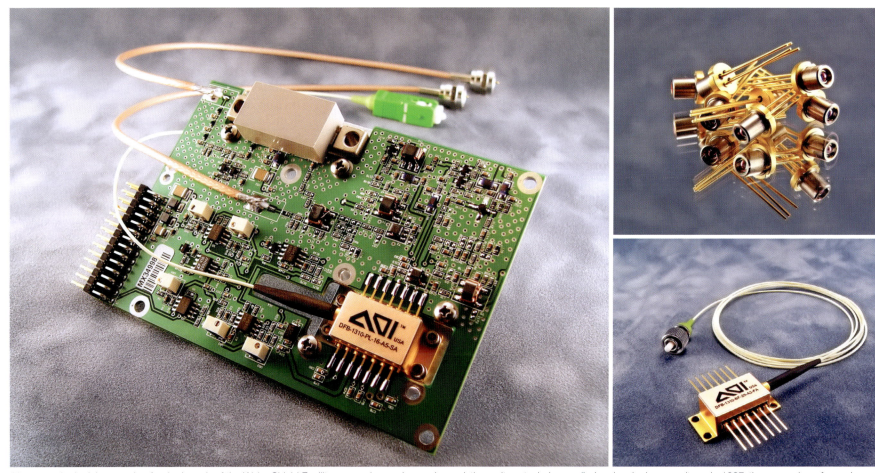

After years in the lab preparing for the launch of the Wake Shield Facility, researchers advanced an existing epitaxy technique called molecular beam epitaxy. In 1997, the researchers formed a company called Applied Optoelectronics Incorporated (AOI) to fabricate devices using the advanced techniques and knowledge. Today, AOI develops and manufactures optical devices for fiber optic networks including cable television, wireless Internet, telecommunications, data communications, and fiber-to-the-home applications.

Toolkits Control Motion of Complex Robotics

Originating Technology/NASA Contribution

That space is a hazardous environment for humans is common knowledge. Even beyond the obvious lack of air and gravity, the extreme temperatures and exposure to radiation make the human exploration of space a complicated and risky endeavor. The conditions of space and the space suits required to conduct extravehicular activities add layers of difficulty and danger even to tasks that would be simple on Earth (tightening a bolt, for example). For these reasons, the ability to scout distant celestial bodies and perform maintenance and construction in space without direct human involvement offers significant appeal.

NASA has repeatedly turned to complex robotics for solutions to extend human presence deep into space at reduced risk and cost and to enhance space operations in low Earth orbit. At Johnson Space Center, engineers explore the potential applications of dexterous robots capable of performing tasks like those of an astronaut during extravehicular activities—and even additional ones too delicate or dangerous for human participation. Johnson's Dexterous Robotics Laboratory experiments with a wide spectrum of robot manipulators, such as the Mitsubishi PA-10 and the Robotics Research K-1207i robotic arms. To simplify and enhance the use of these robotic systems, Johnson researchers sought generic control methods that could work effectively across every system.

Partnership

In 2001, Energid Technologies Corporation, of Cambridge, Massachusetts, was contracted to develop robot control technology through **Small Business Innovation Research (SBIR)** agreements with Johnson. Providing the company with its first funding, these Phase I and II SBIR partnerships, allowed Energid to build unique robot control and simulation software that is now the basis of the company's operations and a solution for a host of robotics applications. The company has also received SBIR funding from NASA's Jet Propulsion Laboratory for additional work related to robotics.

Product Outcome

The human arm is a marvel of engineering; with three joints and seven degrees of freedom (the ability to pivot in seven different ways), the arm's main purpose is to move the hand, allowing a person to grip and manipulate objects in any number of ways and from any number of angles. A "kinematically redundant" robotic arm with seven degrees or more of freedom is designed for the same purpose: to move its hand or other end effector to accomplish tasks. But without the benefit of the natural programming of the human nervous system that enables a person to pick up a pen or turn a doorknob with hardly a conscious thought, a robot arm's velocity, force, direction, and other variables must be carefully controlled through complex algorithms. Energid's patented Actin software, derived from the company's NASA SBIR work, essentially provides a nervous system for kinematically redundant robotic hardware.

"What our software does is let you think only about the hand while we take care of the robot," explains James English, Energid's president and chief technical officer. The Actin toolkit lets the operator put constraints on a robot's motion—such as where the hand should be, how it should move—and then automatically converts those constraints into real-time algorithms to control joint motions for the robot to achieve the desired actions. The result is fluid motion that optimizes key qualities like strength and accuracy while avoiding collisions and joint limits. Actin provides control capabilities for virtually any kind of robot (fixed-base, wheeled, walking), any joint type or tool type, and with any number of joints, degrees of freedom, and bifurcations (branches).

In addition, the software provides machine vision and 3-D visualization tools, as well as powerful simulation capabilities, allowing developers to rapidly devise and test robot designs before the robot is built—a cost- and time-saving measure that results in more efficient and effective robots.

Actin's flexibility across a range of robotic platforms has made the toolkit a go-to solution for multiple cutting-edge robot projects. For the military, Actin provides control capabilities for the U.S. Navy SUMO/FRIEND refueling robotic satellite, in development to extend the lifespan of expensive satellites by keeping them in orbit after their fuel has been exhausted. The software has been adapted to control the U.S. Army's Battlefield Extraction Assist Robot, or BEAR, a powerful lifting robot designed to carry wounded soldiers out of harm's way. Energid has also partnered with iRobot (*Spinoff* 2005) and other companies to develop a robotic bioagent detector based on iRobot's PackBot platform and controlled by the Actin software.

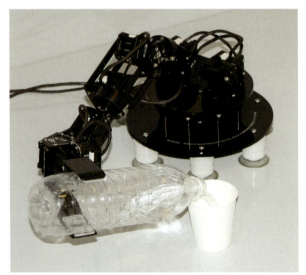

Energid Technologies Corporation's NASA SBIR-derived Actin software provides control of robot manipulators like this Robai Cyton model.

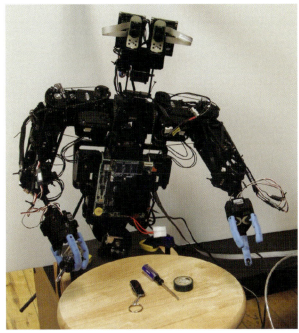

This robot, developed by Energid and its partners, utilizes Actin-based grasping software tools that allow the robot to grasp nearly any object without reprogramming.

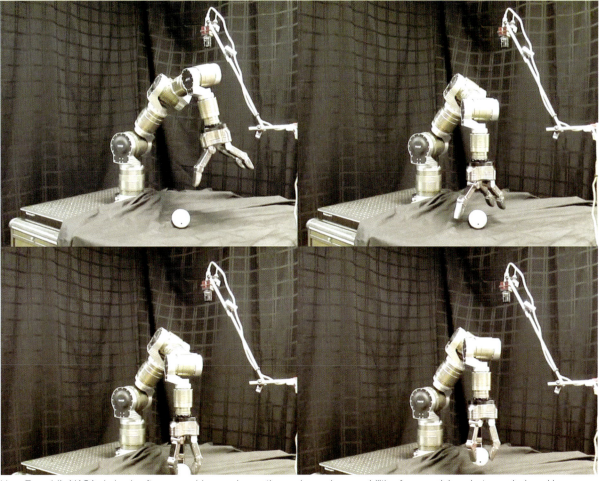

Here Energid's NASA-derived software provides precise motion and grasping capabilities for a modular robot arm designed by German company SCHUNK GmbH & Co. KG. The Actin toolkit allows such robots to make fluid movements with optimized strength and accuracy.

In other areas, Energid's NASA-derived software is at the heart of the company's effort to revolutionize citrus fruit harvesting, using a robotic system with multiple cameras and disposable picking mechanisms to mass harvest fruit for juice processing. Through funding from the National Science Foundation, Energid is developing an Actin-based robotic grasping and manipulation framework that will allow robot manipulators or hands to grasp virtually any object without the need for reprogramming. And through affiliate company Robai, Energid provides its Actin toolkit to control low-cost commercial robot arms.

Energid now has 20 employees and opened an office in India to promote Actin's capabilities in Asia. The foundation of the company's success, English says, lies in the collaboration between government and private industry.

"Someone needs to take the risk to fund the research, and the government is in a good position to do that," he says. "But then someone needs to take the initiative and carry it through to completion, and that is something private industry can do because the motivation is in place. Combining those two is the key." ❖

Actin™ is a trademark of Energid Technologies Corporation.
PackBot® is a registered trademark of iRobot Corporation.

Aeronautics and Space Activities

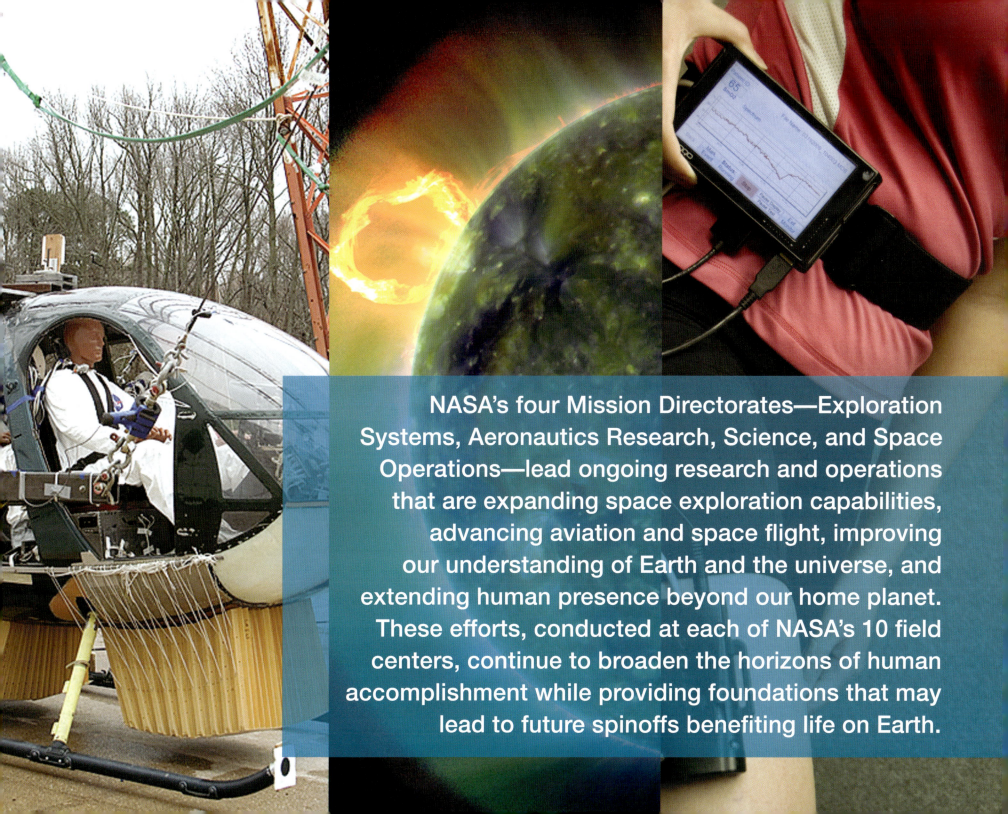

NASA's four Mission Directorates—Exploration Systems, Aeronautics Research, Science, and Space Operations—lead ongoing research and operations that are expanding space exploration capabilities, advancing aviation and space flight, improving our understanding of Earth and the universe, and extending human presence beyond our home planet. These efforts, conducted at each of NASA's 10 field centers, continue to broaden the horizons of human accomplishment while providing foundations that may lead to future spinoffs benefiting life on Earth.

Aeronautics and Space Activities

NASA's mission is to pioneer the future in space exploration, scientific discovery, and aeronautics research. To carry out this mission, the Agency relies upon the ongoing research activities and operational support of its four Mission Directorates: Exploration Systems, Aeronautics Research, Science, and Space Operations. These efforts are conducted at each of NASA's 10 field centers.

Exploration Systems Mission Directorate

NASA's Exploration Systems Mission Directorate (ESMD) develops capabilities and supporting research and technology that will make human and robotic exploration possible. It also makes sure that astronaut explorers are safe, healthy, and can perform their work during long-duration space exploration. In the near-term, ESMD does this by developing robotic precursor missions, human transportation elements, and life-support systems.

Composite Crew Module Designs, Development, Prove Successful

In January of 2007, then-NASA Administrator Dr. Mike Griffin anticipated applications where composite structures could provide benefits to space exploration systems and chartered the NASA Engineering and Safety Center (NESC) to form a team to design, build, and test a full-scale Composite Crew Module (CCM). This effort provided NASA with the experience needed to construct large-scale composite structures using best practices design and production techniques.

To leverage skills and distribute knowledge across the Agency, NESC formed a team with representation from Ames Research Center, Dryden Flight Research Center, Glenn Research Center, Goddard Space Flight Center, Jet Propulsion Laboratory, Johnson Space Center, Kennedy Space Center, Langley Research Center, and Marshall Space Flight Center. Industry partners with varied composite construction expertise included Alcore, ATK, Bally Ribbon Mills, Collier Industries, Genesis Engineering, Janicki Industries, Lockheed Martin, Northrop Grumman, and Tayco. These companies create a wide range of composite products, including aircraft and sailboats. With a mix of skills and cultures, extensive co-locations were established in the first year and nearly 100 percent of the team members were fully dedicated to the project. This intense beginning allowed a smooth transition to a virtual team and the ability to execute a successful project from conception in just over 2 years.

The Composite Crew Module (CCM) sits in the Test Frame in the Combined Loads Test System facility at Langley Research Center.

One of the team's goals was to develop a concept that took complete advantage of the strengths of composite materials. In order to keep launch vehicle mass as low as possible, a new structure design was created. Multiple concepts were brainstormed, and the final CCM became a combination of many features.

The design took advantage of state-of-the-art co-bonded joints that were developed under a contract for

> *Efforts to save weight led to the most complex layup arrangement that the experienced team members, some with B-2 bomber development experience, had ever seen.*

the U.S. Air Force. Three-dimensional woven, pi-shaped, preformed joints were extensively used in place of bonded and bolted joints. Using a building block approach, many specimens with these joints were constructed and tested. The pi preform technology exhibited approximately twice the pull-off strength of traditional L-clip joints.

A unique idea included using a lobed floor with an attached backbone structure. The lobed floor allowed the composite material to be very thin and still carry the internal pressure loads. This saved approximately 50 pounds of mass. Attaching the backbone structure to the floor created a potential structural load sharing with the heat shield during landing. It was estimated that this feature would save approximately 1,000 pounds in heat shield mass.

Tradeoffs between the design, analysis, and fabrication teams were necessary. The CCM sizing, fiber layouts, and analysis were performed using analytical tools such as FEMAP/PATRAN, MSC and NX/NASTRAN, Pro-Mechanica, ANSYS, HyperSizer, LS-DYNA, Thermal Desktop/RadCAD/SINDA, and Fiber SIM. More than 15 finite element models were used to analyze the structure. Efforts to save weight led to the most complex layup arrangement that the experienced team members, some with B-2 bomber development experience, had ever seen. The overall CCM system design was performed in about 12 months, and the analysis effort is running through the entire project.

To manufacture a large-scale composite structure, large-scale tooling, skilled technicians, and extreme attention to detail are required. The tools were constructed with a new technology using chopped fibers impregnated with resin to form the tooling surface. The final dimensions were then machined onto the surface with high-precision mills. This allowed the CCM to be built with dimensional accuracies of 0.010 inch for critical areas and 0.030 inch for noncritical areas.

The overarching process for manufacturing the CCM required first laying up the composite material for the inner skin on the tool and curing it in an autoclave. Next, adhesive and an aluminum honeycomb core were applied and run through an oven cure. Finally, the outer skin was applied and cured in an autoclave.

The CCM was manufactured in two pieces: an upper half and a lower half. This two-piece construction was

It took approximately 5 months for 50 people to fabricate the CCM, which was manufactured in 2 pieces: an upper half and a lower half.

designed to allow more people to be able to work simultaneously on loading instrumentation and equipment into a crew module. The two were later spliced together using an out-of-autoclave process. It took approximately 5 months for 50 people to fabricate the CCM, including the time allowed for the fabrication technicians to train engineers, analysts, and even project managers to perform some of the material layups.

Testing was also a complex process since pressure and combined loads were applied simultaneously. Testing involved designing and building a self-reacting load frame, using many types of instrumentation; impacting the CCM to test for damage tolerance, using different types of non-destructive examination (NDE) techniques; and finding a test facility that could withstand the blast energy when the article was tested to hydrostatic failure.

The reaction load frame was a large steel structure that encompassed the CCM. It allowed the test to be set up with minor modifications to the test facility. The CCM onboard instrumentation included 300 traditional strain gauges, 3,572 fiber optic strain gauges, 8 full-field photogrammetry zones, 2 independent acoustic emission systems, and 2 piezoelectric active acoustic wave monitoring systems.

To perform the damage tolerance tests, critical areas were impacted and then inspected with NDE. Life cycle loads were then applied. No detrimental growth after cycling with damage was detected. NDE techniques were used for monitoring the health of the structure throughout testing, including thorough transmission ultrasonic testing, flash thermography, and visual inspections. The Combined Loads Testing System Facility at Langley was used to perform the tests. It is a large facility with a concrete walled test chamber specifically designed for testing large structures to failure.

CCM technical success criteria included specification that the Preliminary Design Review (PDR) predicted mass should envelope the as-built mass and that the pretest analytical strain predictions should be within

20 percent of the strains measured during load tests. Based on the March 2007 loads, environments, and interfaces, the predicted PDR mass was 1,441 pounds. The CCM as-built mass, including onboard instrumentation, was measured to be 1,496 pounds, within approximately 5 pounds of the PDR prediction. All planned correlation test data strains were approximately 5 percent of the predicted values, well within the 20-percent goal. By paying close attention to details in design and manufacturing, the mass and strength of large composite structures were accurately modeled and predicted.

Using people from multiple NASA centers and multiple industry partners has helped to develop a composites experience network that is rapidly growing throughout NASA. Complex design and build efforts like the CCM increases knowledge in areas such as material systems, damage tolerance testing, analysis methodologies, design and drawing approach, test approach, and non-destructive examination. The CCM team has learned valuable lessons about how to make, inspect, and repair composite structures. This effort will help NASA learn about the benefits of using composite construction as well as help to determine if composite structures will be used for future space exploration systems.

LRO and LCROSS Missions Meet Science Goals

The Lunar Reconnaissance Orbiter (LRO) and Lunar Crater Observation and Sensing Satellite (LCROSS) missions launched from Kennedy on June 18, 2009. After launch, and LRO separation, the LCROSS shepherding spacecraft and the Atlas V's Centaur upper stage rocket executed a flyby of the Moon and entered into an elongated Earth orbit to position LCROSS for impact at the lunar South Pole.

LRO completed its commissioning phase on September 15, 2009, and its observations will be used to identify potential safe landing sites for astronauts, to locate potential resources, to characterize the radiation environment, and to demonstrate new technology. The LRO spacecraft remained in low polar orbit for its 1-year exploration mission, and returned global data sets that will be used to create temperature maps, a global geodetic grid, and high-resolution color imaging of the entire lunar surface. This mission places particular emphasis on the polar regions of the Moon where continuous access to solar illumination may be possible and in the permanently shadowed regions at the poles where water exists.

On October 9, 2009, LCROSS completed its mission to confirm the presence or absence of water ice in the permanently shadowed Cabeus crater at the Moon's South Pole. After separating from the LCROSS shepherding spacecraft, the Centaur became an impacting vehicle, creating a debris plume that rose above the lunar surface. Following 4 minutes behind, the shepherding spacecraft flew through the debris plume, collected and relayed data back to Earth before also impacting the lunar surface. This second impact ejected material from the crater's surface that created a plume of water (ice and vapor), hydrocarbons, and hydrated materials.

LCROSS' science goals were fully met. The Centaur impacted the targeted area in the Cabeus crater within 100 meters. The plume rose approximately 5 kilometers, and even though not visible from Earth, was clearly detected by instruments on both LCROSS and LRO.

The Atlas V/Centaur rocket, the Lunar Reconnaissance Orbiter (LRO), and the Lunar Crater Observation and Sensing Satellite (LCROSS) wait at Launch Complex-41 at Cape Canaveral Air Force Station in Florida.

LRO spacecraft data sets will be used to create temperature maps, a global geodetic grid, and high-resolution color imaging of the entire lunar surface.

Prototype Sensor Measures Blood and Tissue Chemistry Without Incision

The National Space Biomedical Research Institute (NSBRI) is funding new research and technology to develop a system called the Venus prototype, under the direction of Dr. Babs Soller from the University of Massachusetts Medical School in Worcester, Massachusetts. The Venus is a medical technology that will enable a sensor system to measure a person's blood and tissue chemistry with no need for painful incisions or blood draws. The noninvasive, needle-free system uses light to measure tissue oxygen and potentiometric hydrogen ion concentration (pH, a measure of acidity or basicity).

This NASA-funded technology may eventually serve as an alternative to drawing blood, without the use of additional medical equipment, to determine a person's metabolic rate. The Venus has the ability to measure blood, tissue chemistry, metabolic rate (oxygen consumption) and other parameters.

To take measurements, the Venus prototype is placed directly on the skin. The 4-inch by 2-inch sensor uses near-infrared light (light just beyond the visible spectrum), and while some blood in the tiny blood vessels absorbs some of the light, the rest is reflected back to the sensor. The monitor then analyzes the reflected light to determine metabolic rate, tissue oxygen, and pH.

Soller, who leads the NSBRI Smart Medical Systems and Technology team, explains how the technology can be used in space and on Earth. "The measurement of metabolic rate will let astronauts know how quickly they are using up the oxygen in their life-support backpacks. Tissue and blood chemistry measurements can also be used in medical care to assess patients with traumatic injuries and those at risk for cardiovascular collapse."

Risk and Knowledge Management System Implements Safety, Reliability

A new system was initiated in 2006 at ESMD called the Integrated Risk and Knowledge Management (IRKM) system. The foundation of this system is continuous risk management (CRM).

CRM requires an evaluation of events coupled with proactive measures to control or mitigate risks. A novel aspect of the IRKM approach is that it uses risk records resulting from the CRM process to initiate an assessment of what information to transfer to solve a problem. It then follows up to capture the actual strategy or measures used to mitigate the risk. Risk records used in this fashion provide a cueing function similar to an aircraft sensor cueing a weapons system sensor. In the IRKM system, CRM informs knowledge management, and knowledge management becomes the enabler of CRM.

CRM identifies, analyzes, plans, tracks, controls, communicates, and documents risk through all life cycle phases of an organization's product development. ESMD uses an enterprise risk management approach and a common framework for identifying, analyzing, communicating, and managing risks. Risks are communicated vertically through a well-defined escalation process, while horizontal integration occurs through a multi-tiered working group and board structure. This network of risk managers is also used to communicate lessons learned and best practices—referred to as a central nervous system for information flow.

The IRKM system has an important work-process-assist element called Process 2.0, or P2.0, which is in part modeled after the U.S. Army after-action review process. P2.0s are process-focused, collegial, structured reflection events. There has been huge demand for the P2.0 events, which assist teams in examining all aspects of a given process, including stakeholders, inputs, outputs, and products. P2.0 events use critical process mapping, structured brainstorming techniques, and process failure modes and effects analysis to identify and address process issues.

As an option, P2.0 users can take advantage of a Web-based collaboration tool suite. The tool provides a simple-to-use information capture capability that increases the volume and speed of idea capture and supports alternative analysis. Most important, the P2.0 method demands disciplined thinking to drive out process improvements for the team. P2.0s have been used to assist a diverse set of team processes, ranging from vibro-acoustic coupled-loads analysis to a simple integration meeting gone awry. In each case, the result has been rapid, transparent, team-authored process improvement.

ESMD risk records provide the context for knowledge-based risks—Web-based, multimedia knowledge bundles that provide users with expert advice on risk control and mitigation for specific technical risks. ESMD defines a knowledge-based risk as a risk record, with associated knowledge artifacts, to provide a storytelling narrative of how the risk was mitigated and what worked or did not work. As key risks are mitigated, particularly risks that are likely to recur across other programs in ESMD, knowledge is captured and transferred. Knowledge-based risks identify the effectiveness of mitigation activities, specifically in terms of cost, schedule, and technical performance. Instead of a collect, store, and ignore approach, knowledge-based risks form an active collection of lessons learned that are continually reused and updated.

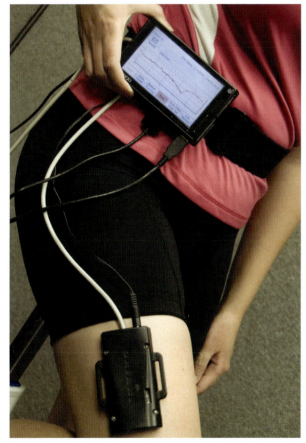

The Venus prototype is a noninvasive, needle-free system that uses light to measure tissue oxygen and potentiometric hydrogen ion concentration (abbreviated as pH). It consists of a sensor (shown on the thigh) and a wearable monitor (shown on the waist).

The ESMD wiki environment enables horizontal communication, collaboration, and knowledge sharing across ESMD. More than 350 wikis provide a multi-functional toolset to assist ESMD teams. An important part of exploiting the wiki technology has been helping teams critically examine their work processes and information architecture, which is then mapped into the tool. The wiki provides teams with an easy-to-use, flexible interface to collaborate on documents, conduct discussions, manage calendars, locate information, and, most important, work more effectively.

Knowledge capture and transfer activities are designed to document project execution lessons learned and best practices using a conversation-based format. While overlapping in some respects, knowledge capture and transfer differs from P2.0 in that it focuses on project execution rather than recurrent process implementation. Knowledge capture and transfer rejects the notion of asking participants to fill out questionnaires. Rather, knowledge capture and transfer uses the most natural modality—conversation, but carefully structured and controlled conversation. Project risk records are used to guide the initial interviews. Individual issues are synopsized and aggregated, and a composite analysis is provided. Results are rapidly provided to stakeholders using a variety of communication modes, including briefings, design review checklists, peer assists, knowledge cafes (small group brainstorming), and video interviews.

The Riskapedia wiki space is intended to assist ESMD programs, projects, managers, and workers in implementing life cycle risk management practices and discipline. Riskapedia provides extensive content (tools, techniques, best practices, videos, and lessons learned) addressing the fundamental blocking and tackling skills of risk management: risk identification, risk assessment, and risk control and mitigation planning. The resource is a hard hat area that is intended to be under construction for life. The space has been populated with expert-developed content that is intended to evolve over time as users and contributing editors engage in ongoing construction of subject matter articles. Users have the opportunity to rate and discuss content, provide or author content (as a contributing editor), ask questions of experts, and use content in the performance of work and the management of risks.

ESMD risk records illuminate top engineering management and technical issues. Each case is structured to highlight key transferrable aspects of risk management. The proper application of risk management principles can help manage life cycle costs, development schedules, and technical scope, resulting in safer and more reliable systems for NASA's future programs. Examining the critical thinking that made past programs successful could enhance the technical curiosity of engineers developing future space systems and make the programs equally robust.

Aeronautics Research Mission Directorate

NASA's Aeronautics Research Mission Directorate (ARMD) conducts cutting-edge, fundamental research in traditional and emerging disciplines to help transform the Nation's air transportation system and to support future air and space vehicles. Its goals are to improve airspace capacity and mobility, improve aviation safety, and improve aircraft performance while reducing noise, emissions, and fuel burn.

Ductile Superalloy Disk Coating Extends Life of Engine Components

ARMD worked with GE Aviation to develop a ductile metallic coating to protect engine components from damage. The coating is now being further developed by the Naval Air Warfare Center for potential application as corrosion protection for turbine disks. It is also being incorporated by the U.S. Air Force in the Hybrid Disk Program for high-temperature, extended-duration disk applications.

Powder metallurgy superalloy high-pressure compressor and turbine disks and seals in engines can experience time-dependent damage at exposed surfaces. Tests show fatigue life can be reduced up to 80 percent by static and cyclic exposures at disk operating temperatures even in lab air, through activation of surface crack initiation mechanisms. High-pressure turbine disks and seals in current engines can also experience hot corrosion-related pitting. The effects of hot corrosion damage in the form of pits have been found to even more severely reduce fatigue life, by up to 98 percent in accelerated corrosion test conditions.

Superalloy turbine blades, which are exposed to higher temperatures but reduced stresses, have been protected by metallic and ceramic coatings to prevent such forms of surface attack. The associated disks that hold all of the blades, however, are subjected to lower temperatures but much higher fatigue stresses, which have previously precluded such coating protection. Increasing temperatures and service times for advanced engines in the field were causing surface damage, and a coating was needed that could extend the life of exposed disk surfaces, yet not harm the inherent fatigue resistance of the superalloy.

Tests show fatigue life can be reduced up to 80 percent.

A series of experimental coating alloys based on nickel with chromium and aluminum additions were screened in accelerated hot corrosion tests to ensure the coating could prevent formation of corrosion pits for long periods of time. The alloys were simultaneously applied to fatigue specimens and tested to examine the effect of the coating on fatigue resistance. A suitable alloy was found to extend corrosion life by over 200 percent, yet did not impair disk fatigue resistance. Fatigue specimens tested after extended periods of corrosion attack still did not fail from the exposed surfaces. The coating was found to be both adherent and resilient after water quenching, bending, and burner rig testing.

NASA continues to work with GE Aviation to test and model the benefits of the coating on disk life, in oxidation and hot corrosion conditions at current and higher disk temperatures.

High-Speed, Non-Immersion, Ultrasonic Scanning System Decreases Inspection Time

Early damage detection is critical for safe operation and cost-effective maintenance of aircraft structures. Traditional ultrasonic inspection can be an effective health assessment method, but current non-immersion methods can be slow and difficult to operate in the field. To implement a relatively low-cost, high-speed, and high-resolution ultrasonic scanning system that was simple and easy to operate, NASA researchers combined an improved non-immersion ultrasound method with a unique motion synchronization technique.

The system core consists of an open-loop motion control and data acquisition platform. Using commercial-off-the-shelf hardware and custom multi-threaded control software, the system acquires time-domain data signals in precise synchronization with a continuous scanning motion. The synchronization method allows a low-cost stepper motor controller to generate high-speed motion synchronized output, which triggers the internal digitizer and any external excitation equipment. This yields a flexible platform suitable for any nondestructive evaluation (NDE) method requiring motion synchronized time-domain acquisition such as eddy current, acoustic, and ultrasound. System software provides a live C-Scan display while simultaneously recording time-domain waveforms or A-Scan at every inspection point.

This non-immersion method uses a captured water column with a durable membrane to couple sound from a standard (or custom) immersion probe. The inspector can select transducer frequency, geometry, and focusing characteristics to meet the inspection requirements, as one would in an immersion inspection. The durable membrane can be adjusted to accommodate a wide range of rough or uneven surfaces, and the method does not require a water pump, vacuum, hoses, or complicated seals. The operator need only apply a light mist of soapy water to promote acoustic coupling and decrease wear. The membrane is easy to replace and the water column is easily set up with a syringe.

The current implementation can be mounted on a range of different load frames for in-situ high-resolution inspection of specimens under test. However, the approach can also be widely applied to scanning subsystems in the field or laboratory. Existing scanning systems in the NASA NDE laboratory will be refurbished using this technique. The new system can decrease defect thresholds, inspection cost and time, and increase inspection area and frequency.

Subscale Flight Testing Facility Provides Proving Ground for Flight Technologies

As part of a long-term strategy to improve aviation safety, engineers at NASA are looking for ways to avoid, and if necessary, mitigate, loss-of-control events in transport aircraft. Achieving this strategy requires an understanding of the aircraft's dynamics in abnormal and upset flight conditions. This is when events such as structural damage, hydraulic failures, or ice buildup change the vehicle's performance to an extent that traditional autopilots are unable to fly the aircraft and pilots may be faced with highly coupled controls and oscillatory or even divergent handing characteristics.

The development of assistive control technologies requires a rich set of experimental data, obtained near the boundaries of the operating envelope. This is where the Airborne Subscale Transport Aircraft Research (AirSTAR) facility comes into play. AirSTAR is both a ground facility and a set of unmanned aircraft. The ground facility is similar to a modern flight simulator, with computer-generated out-the-window views, detailed instrumentation overlays on a heads-up display, and a full set of side-stick, throttles, and rudder pedals for pilot interface.

The unmanned vehicles in the Airborne Subscale Transport Aircraft Research facility (left) are only about 1/20th the size of the aircraft they are built to represent. Inside the mobile operations station (right) is a flight computer and research pilot station with several displays, including a navigation display and an analog video feed from a nose-mounted camera.

The difference between AirSTAR and a flight simulator is that in AirSTAR, the vehicle is actually flying as a remotely piloted aircraft communicating with display and control computers over a high-bandwidth telemetry link. An openly distributable simulation model has also been developed for AirSTAR, which allows advanced control algorithms to be quickly evolved from proof-of-concept simulations through validation flight tests.

> *The unmanned vehicles in the AirSTAR facility are only about 1/20th the size of the transport aircraft they are built to represent.*

The unmanned vehicles in the AirSTAR facility are only about 1/20th the size of the transport aircraft they are built to represent. With a careful structural design that scales the mass distribution and density along with the geometry, these sub-scale flight vehicles provide flight-dynamic and aerodynamic responses that are faster, but otherwise identical, to their full-scale counterparts. This retains relevance to the target application, and allows experiments to be conducted that have more risk and incur larger structural loads than would be feasible on a full-scale transport.

In September 2009, the AirSTAR system reached operational status and flew from an unmanned vehicle runway at NASA's Wallops Flight Facility. One of the experiments performed included automated surface perturbations during a slow approach to stall, which allowed for the identification of flight models at high angles of attack. The flights also included simulated damage to the elevator and degradation in aircraft stability. Fully adaptive control algorithms were implemented and tested with pilots evaluating the benefits of these controller-assisted flight modes under emergency conditions. These test results, and future high-risk modeling and controls experiments planned with this facility, provide a proving ground for flight technologies that will continue to improve aviation safety in the next generation of transport aircraft.

Estimation for Health Management Technologies Reduces Uncertainties

Aircraft engines are highly complex systems consisting of static and rotating components, along with associated subsystems, controls, and accessories. They are required to provide reliable power generation over thousands of flight cycles while being subjected to a broad range of operating conditions, including harsh temperature environments. Over repeated flight cycles, engine performance will degrade and engine malfunctions may occur.

Under the Aviation Safety Program's Integrated Vehicle Health Management (IVHM) project, NASA researchers are developing innovative Estimation for Health Management (EHM) technologies to assist aircraft operators in managing the safety and reliability of their gas turbine engine assets. This includes the development of real-time onboard models for the in-flight estimation of engine performance parameters that can be directly utilized by EHM applications. These onboard models are designed with the ability to self-tune, or adapt, based on sensed measurements to track the performance of the actual engine. A challenge associated with developing accurate onboard models is the fact that engine components will naturally experience degradation over time, affecting an aircraft engine's performance.

The level of engine degradation is generally described in terms of unmeasured health parameters like the efficiency and flow capacity of each major engine component. Using mathematical estimation techniques, health parameters and the level of performance degradation can be estimated, given that there are at least as many sensors as parameters to be estimated. In an aircraft engine, however, the number of sensors available is typically less than the number of health parameters, presenting an underdetermined estimation problem. A common approach to address this shortcoming is to estimate and adjust a subset of the model health parameters, referred to as model tuning parameters. This approach will enable the onboard model to track measured engine outputs. Model-produced estimates of unmeasured engine outputs may be inaccurate, however, due to the fact that the impact of all the health parameters will not be accurately represented within the model.

To address this challenge, NASA has developed an innovative methodology that constructs a set of model tuning parameters defined as a linear combination of all health parameters and of appropriate dimension to enable estimation. Selection of the tuning parameters is performed using an analytical method designed to minimize the estimation error in the model parameters of interest. The new methodology has been validated in simulation using an aircraft turbofan engine simulation. The results demonstrated that applying the enhanced tuning parameter selection methodology resulted in a 30-percent reduction in average estimation error compared to the conventional approach of selecting a subset of health parameters to serve as the tuning parameters.

This technology holds great potential for applications that require a real-time estimate of unmeasured engine outputs. It is not practical to sense every parameter of interest due to cost, weight, and harsh high-temperature environment constraints. Therefore, parameter estimates are often indirectly inferred through other sensed measurements. Improving the accuracy of these synthesized parameter estimates through onboard adaptive models can help reduce uncertainty, and directly improve applications such as engine diagnostics, controls, and usage-based life consumption calculations.

Southwest Airlines Identifies Anomalies Using NASA's Data Mining Tool

Southwest Airlines currently analyzes data for about 1,600 flights each day from 305 different aircraft. It uses a third-party analysis tool to identify threshold exceedances based on flight operations manual limits and other coordinated parameter limits. A daily review and voluntary pilot reporting are how most anomalies are discovered. However, anomalies in multivariate datasets are often represented by more than just single-variable exceedances. Individual variables may be within normal ranges while the normal relationships among them may be violated.

Orca is one of several multivariate anomaly detection methods developed by researchers at Ames Research Center. Southwest Airlines experimented with Orca on 7,200 flight segments containing descents from 10,000 feet to touchdown on a single runway. Orca identified significant anomalies such as high roll and pitch events near the final approach and hard nose-over events prior to landing. These results have caused Southwest to add new events to their daily exceedance review, which is performed by the third-party analysis tool to determine threshold exceedances. NASA and Southwest plan to continue exploring the use of various NASA-developed and other data mining methods for anomaly detection on increasing amounts of flight data.

The goal of NASA's IVHM project is to develop validated tools, technologies, and techniques for automated detection, diagnosis, and prognosis that enable mitigation of adverse events during flight. Indications of these adverse events are thought to manifest themselves within the vast amount of data produced during flight by the thousands of aircraft and associated systems and people that operate in the skies each day. Fortunately, the number of adverse events experienced is very small relative to the amount of activity occurring. Detecting the small subsets of the data that represent adverse events is the first step to determining exactly what went wrong and identifying the most probable precursors of an adverse event.

A significant area of research and development in NASA and other organizations is in model-based methods for fault detection. Such methods involve developing a model that represents normal operation of a complex engineered system such as an aircraft, and identifies sensed data that are significantly different from what

NASA tools identified significant anomalies in flight data, allowing Southwest Airlines to add new events to their daily review. NASA and Southwest plan to continue exploring the use of various NASA-developed and other data mining methods for anomaly detection on increasing amounts of flight data.

the model predicts. Such significantly different data are assumed to represent abnormal operation. These methods are often effective because the models that they use represent domain expert knowledge. However, these models are typically unable to represent the full range of normal operating modes including those that represent small amounts of degradation, which are not enough to be considered adverse events. Much of this knowledge is contained within the vast amount of data representing many years of operation of many different aircraft. Data mining methods can extract this knowledge.

Data mining methods for anomaly detection assume that most of the training data supplied represent normal operations, so the methods try to find the small amount of data that is significantly different from the rest. Research and development in this area aim to find anomalies of various types in vast data repositories. Examples include strange sequences of computer system commands that

Test instrumentation is set up behind the inboard engines of NASA's DC-8 airborne science laboratory during alternative fuels emissions and performance testing at NASA's Dryden Aircraft Operations Facility in Palmdale, California.

may represent hackers' activities, unusual weather events, or unusual aircraft operations.

Researchers at Ames have developed several anomaly detection methods including Orca, which is designed to quickly find distance-based anomalies. Distance-based anomaly detection methods flag data points that are furthest from most of the data, which is assumed to be normal. In Orca, the user can set a distance metric to use, such as the average distance to a point's k-nearest neighbors for some user-chosen number k, and the maximum number, N, of anomalies to be identified. Orca then finds the N points that have the highest value for that distance metric. Orca does this efficiently by typically not requiring the comparison of all pairs of data points.

Data-driven anomaly detection methods can be used for domains other than commercial aviation. Researchers at Ames have used these methods to identify anomalies in climate data, space shuttle main engine test data, and space station operations, among others. NASA is continuing research into making novel anomaly detection methods that are fast enough to analyze very large repositories of data, with applications to the many problems of interest to NASA. They are also working on extending these methods to heterogeneous data—data that do not just include continuous measurements, but also include discrete data (such as pilot switches and system modes), and even text repositories (such as reports written by flight crews and passengers that describe problems that occurred during flight). The goal is to develop methods that automatically analyze these data to not only detect anomalies, but also to identify what happened to bring about these anomalies, as well as why it happened. This information will enable mitigation of adverse events and improve safety for the flying public.

Alternative Fuels Tested to Potentially Provide Aviation Fuel

The high cost of aviation fuel, the growing demand for air travel, and the environmental issues associated with petroleum fuels, have prompted the aviation industry and engine companies to

Some alternative fuels and fuel blends show promise as aviation fuels.

Langley Research Center's Bruce Anderson and United Technologies' David Liscinsky install tubing to connect pressure ports located on the exhaust inlet probe with sensors located in equipment trailers for the Alternative Aviation Fuel Experiment (AAFEX).

look for alternative sources of aviation fuel. Some alternative fuels and fuel blends, including renewable biofuels, are appealing alternatives to current hydrocarbon-based fuels and show promise as aviation fuels.

To determine the effects of future alternative fuels on emissions from aircraft engines, NASA partnered with the U.S. Department of Defense, U.S. Environmental Protection Agency (EPA), and Federal Aviation Administration to conduct an experiment examining the performance and emissions of alternative fuels using the NASA Dryden Flight Research Center's DC-8 aircraft.

In early 2009, a team of NASA engineers and 11 other research groups conducted specific field tests to assess combustion emissions, engine performance characteristics, and engine exhaust plume composition of 2 non-petroleum-based jet fuels. The Alternative Aviation Fuel Experiment (AAFEX) tests were conducted in late January through February 3 at Dryden using synthetic fuels produced with the Fischer-Tropsch (F-T) process. One was derived from natural gas and the other was produced from coal. Both had drawn attention because of their potential for the high energy necessary for commercial flight and they were both available in the quantities required for this large-scale test.

The F-T process is a chemical reaction in which a synthesis gas (a mixture of carbon monoxide and hydrogen) is converted into hydrocarbons of various forms. The process can produce synthetic petroleum for use as a lubricant or fuel, and has been around for decades. Until now, the high cost of building new plants to produce synthetic fuels has stymied interest, except in South Africa, where Sasol, an energy and chemicals company, has been producing jet fuel from coal for a number of years. The United States does not have any F-T plants, although synthetic fuel is now being produced using this process at a number of other locations around the world.

The DC-8 was utilized for the experiments because it had been used previously for an extensive series of emissions testing in the Aircraft Particle Emissions Experiment (APEX). For each of the two F-T fuels, researchers tested both 100-percent synthetic and 50-50 blends of synthetic and standard jet fuel. Almost all previous engine testing had considered only 50-50 blends, and no aircraft had been previously tested using 100-percent synthetic fuel.

Researchers found that burning F-T fuel did not appreciably affect engine performance, but did lead to aircraft and storage tanker fuel leaks for the pure F-T fuels due to seal shrinkage from exposure to the aromatic-free fuels. Small effects of synthetic fuel were found on NOx, CO, and unburned hydrocarbon emissions. The most profound effect of the synthetic fuels, however, was to reduce engine black carbon number density and mass emissions by as much as 75 percent, relative to JP-8, at lower power conditions. Particulates were reduced across the full range of engine powers but reductions were less at higher power. The F-T fuels also reduced hazardous

Harvard University graduate student Ben Lee tunes the quantum-cascade-laser methane isotope sensor for the AAFEX tests.

The electron beam freeform fabrication (EBF3) process could provide a way for astronauts to fabricate structural spare parts and new tools aboard the International Space Station.

air pollutant emissions, and the fuel's lack of sulfur impurities suppressed formation of volatile aerosols in the test engine's exhaust plume as it cooled and mixed with ambient air.

A limited amount of emissions testing was also conducted using the auxiliary power unit of the aircraft with one of the pure F-T fuels and standard jet fuel. Engine performance, when operating on the F-T fuel, was not appreciably affected, though particulates were substantially reduced.

In addition to collecting needed emissions data for the fuels, NASA is ensuring that the other information is available to the community by developing a database for alternative fuels. The database was created using Microsoft Access and currently has all the standard properties specified in the current ASTM D1655 (a standard specification for aviation turbine fuels) for 19 different synthetic fuels and fuel blends. The database can provide reports for selected fuels, fuel properties, or ranges of fuel properties that can be printed or exported to an Excel file for further processing. Additional fuels can be added to the database as data becomes available.

NASA is one of many organizations aggressively working to understand how non-petroleum alternatives may be used to satisfy the growing demand for less expensive, cleaner burning fuel for aviation. The AAFEX tests were funded and managed by the NASA Fundamental Aeronautics Program (FAP) of ARMD.

FAP is continuing to conduct research and is making significant progress in advancing the alternative fuels technology. The majority of this work is primarily conducted under the Subsonic Fixed Wing Project, and capitalizes on NASA's technical expertise and relevant facilities to increase our knowledge of fuel characterization and performance. Other related work includes the measurement of thermal stability of alternative fuels and fuel blends to determine suitability for aircraft use and fundamental studies of the F-T reaction and catalyst effects on aviation fuel yield. Research is also being conducted in the area of renewable biojet fuels and includes fundamental studies on biomass for efficient jet fuel production as well as emissions testing of biojet fuels. A second alternative fuel test is planned for the NASA DC-8 using renewable biojet fuels in 2011.

Electron Beam Fabrication Revolutionizes Aircraft Designs and Spacecraft Supportability

Electron Beam Freeform Fabrication (EBF3) is a cross-cutting technology advance in layered part fabrication for producing structural metal parts. Developed by researchers at NASA's Langley Research Center as a replacement for forgings, this manufacturing process offers significant reductions in cost and lead time. The promise of this technology extends far beyond its applicability to low-

This structural metal part was fabricated from EBF3, an advancement in layered part fabrication.

cost manufacturing and aircraft structural designs. EBF3 could provide a way for astronauts to fabricate structural spare parts and new tools aboard the International Space Station (ISS) or on the surface of the Moon or Mars.

EBF3 uses a focused electron beam in a vacuum environment to create a molten pool on a metallic substrate. EBF3 works in a vacuum chamber, where an electron beam is focused on a source of metal, which is melted and then applied as called for by a drawing—one layer at a time—on top of a programmed moving surface until the part is complete. EBF3 has two key requirements: A detailed three-dimensional drawing of the object to be created, and a material that is compatible for use with an electron beam.

The drawing is needed to divide the object into layers, with each cross-section used to guide the electron beam and source of metal in reproducing the object, building it up layer by layer. The material must be compatible with the electron beam so that it can be heated by the stream of energy and briefly turned into liquid form (aluminum is an ideal material). EBF3 can actually handle two different sources of metal—also called feed stock—at the same time, either by mixing them together in a unique alloy or embedding one material inside another. The potential use for the latter could include embedding a strand of fiber optic glass inside an aluminum part, enabling the placement of sensors in areas that were previously impossible.

This layer-additive process enables fabrication of parts directly from computer-aided design drawings. Metal can be placed only where it is needed and the material chemistry and properties can be tailored throughout a single-piece structure, leading to new design methods for integrated sensors, tailored structures, and complex, curvilinear (characterized by following a curved line) stiffeners. The parts can be designed to support loads and perform other functions such as aeroelastic tailoring or acoustic dampening.

Future lunar-based crews could use EBF3 to manufacture spare parts as needed, rather than rely on a supply of parts launched from Earth. Astronauts could possibly mine feed stock from the lunar soil, or even recycle used landing craft stages by melting them. The immediate and greatest potential for the process, however, is in the aviation industry where major structural segments of an airliner, or casings for a jet engine, could be manufactured for about $1,000 per pound less than conventional means. Environmental savings also are made possible by deploying EBF3.

To demonstrate the potential of EBF3 for on-orbit applications, an ISS EBF3 demonstration to produce on-demand enabling tools, parts, and structures, is being proposed. The theme of the proposed demonstration is "Materials Science and Supportability." On-demand, additive manufacturing mitigates spare parts mass and volume requirements to increase payload capability on long-duration missions, and also enhances mission flexibility. EBF3 also enables repair and assembly of large structures on-orbit. The main objective is to advance the technology readiness level of on-demand manufacturing and demonstrate enabling environmentally friendly fabrication and repair technology on-orbit. Basic experiments in basic materials science, physics, and dynamics of molten metals in microgravity will help to mature the technology.

The maturation of the EBF3 technology from inception to commercialization, including the formation of a government-industry team to complete the handoff to the industrial manufacturing sector, will exemplify a significant technology spinoff to industry that continues to forge a stronger partnership between NASA and industry. It will also describe how a manufacturing process can influence future aircraft designs by providing a solution that enables multidisciplinary optimization, and continues to demonstrate the contributions to advance technology and innovation from NASA's research and development. Finally, it will end with the transition from a materials process development activity to a tool that can change design methodologies to incorporate aeroelastic and acoustic tailoring into aircraft structures.

Although developed for aerospace applications, EBF3 advances metal manufacturing techniques and adds new capability to the rapid manufacturing and rapid prototyping industry for manufacturing complex components and customized parts. Spinoff opportunities for on-demand manufacturing of custom or prototype parts exist in industries from custom automotive applications to racing bikes, sporting goods to medical implants, oil drilling tools, and power plant hardware. A small, robust EBF3 system, such as that proposed for demonstration on the ISS, also offers a capability for conducting repairs and building replacement parts in remote locations such as Antarctic scientific bases, Navy ships, military operations, and in space.

Energy Absorbing Technology Tested for Improved Helicopter Safety

According to the National Transportation Safety Board, more than 200 people are injured in helicopter accidents in the United States each year, in part because helicopters fly in riskier conditions than most other aircraft. They fly close to the ground, not far from power lines and other obstacles, and often are used for emergencies such as search and rescue and medical evacuations.

NASA aeronautics researchers recently dropped a small helicopter at the NASA Langley Landing and Impact Research Facility from a height of 35 feet to investigate how effective an expandable honeycomb cushion called a deployable energy absorber (DEA) would be in mitigating the destructive force of a crash. The objective of the drop test with the DEA was to make helicopters safer, under

Researchers at Langley are testing the deployable energy absorber with the help of a helicopter donated by the Army, a crash test dummy contributed by the Applied Physics Laboratory in Laurel, Maryland, and a 240-foot-tall structure once used to teach astronauts how to land on the Moon.

A honeycomb-type airbag that was created to cushion astronauts may end up in helicopters to help prevent injuries.

circumstances of uncontrolled scenarios and environments to prevent severe or even catastrophic injuries to crew and passengers.

For the test, researchers used an MD-500 helicopter donated by the U.S. Army. The rotorcraft was equipped with instruments that collected 160 channels of data. One of the four crash test dummies was a special torso model equipped with simulated internal organs, provided by the Johns Hopkins University Applied Physics Laboratory in Laurel, Maryland. The underside of the helicopter's crew and passenger compartment was outfitted with the DEA, which is made of Kevlar and has a unique flexible hinge design that allows the honeycomb to be packaged flat until needed. The DEA was invented and patented by Dr. Sotiris Kellas, a senior aerospace engineer in the Structural Dynamics Branch at Langley.

The MD-500 helicopter was crash-tested by suspending it about 35 feet into the air using cables, and was then swung to the ground, using pyrotechnics to remove the cables just before the helicopter hit so that it reacted like it would in a real accident. The test conditions imitated what would be a relatively severe helicopter crash. The flight path angle was about 33 degrees and the combined forward and vertical speeds were about 48 feet per second, or 33 miles per hour.

On impact, the helicopter's skid landing gear bent outward, but the cushion attached to its belly kept the rotorcraft's bottom from touching the ground. Data from the four crash test dummies were compared with human injury risk criteria and the results indicated a very low probability of injury for this crash test. In addition, the airframe sustained minor damage to the front right subfloor region. The crash data will be further analyzed to determine whether the DEA worked as designed. The acquired data will be used to validate the integrated computer models for predicting how the different parts of the helicopter and the occupants react in a crash, while the torso model test dummy will help to assess internal injuries to crew and passengers in a helicopter crash.

The damage was repaired and the airframe was used in a second full-scale crash test, this time without the deployable energy absorber. The test went as anticipated and was successful. With the countdown to the release, the helicopter hit the concrete. Its skid gear collapsed, the windscreen cracked open, while the occupants lurched forward violently, suffering potentially spine-crushing injuries according to internal data recorders.

As the results of the crash dynamics of the helicopter and the impact on the crew and passengers (dummies) are better understood, it will provide improved crash performance to enhance safety and minimize the severity of injuries to crew and passengers. Thus, there is also the potential for application of the DEA technology in other aircraft, including commercial aircraft, to save lives in future aircraft accidents as well.

Aviation Safety Research Advances Optical Neuroimaging

Human performance issues are often cited as causal factors of aviation accidents. Even the most expert and conscientious pilots are susceptible to making errors in certain circumstances. During prolonged critical activities, a person's performance can decrease due to fatigue or workload. Continuous monitoring of attention and performance is important for continuous safety. Intelligent cockpits of the future will interact with pilots in ways designed to reduce error-prone states and mitigate dangerous situations at the edges of human performance. One critical aspect is to develop reliable and operationally relevant metrics for the state of the operator using noninvasive, portable, safe, and inexpensive means.

To assist, recent research by NASA Aviation Safety aims to enable functional near-infrared spectroscopy (fNIRS) as a replacement or complementary technique to electroencephalography, or EEG, and other physiological measurements. The fNIRS process is a noninvasive, safe, portable, and inexpensive method for monitoring brain activity. It uses both visible and near-infrared light to sense blood flow in the cortex and quantify changes in concentration of oxygen in the blood indicating neural activity. This research could help to reduce the effects of performance decrement and improve safety by informing intelligent systems of the state of the operator during flight.

The fNIRS method uses light sources placed on the surface of the scalp as well as paired detectors that receive the light as it is returned from the scalp and outer layers of the brain. However, the hardware currently used to obtain optical signals to and from the scalp typically requires a heavy helmet. These difficulties have held the technique back from industry and academia, though fNIRS currently works well in the laboratory.

To address these obstacles, biomedical engineers at Glenn Research Center are developing next-generation headgear concepts for fNIRS. This work could also enable the use of fNIRS for monitoring during any activity that pushes the limits of human performance. Moving away from helmet-based mounting systems, results have shown the optical and ergonomic usefulness of a material that allows the development of headgear to be both comfortable and reliable. The material is a lightweight, cleanable, curable, biomedical-grade elastomer that transmits the light while increasing comfort. Since optical component-to-skin contact is required, existing fNIRS headgear can be painful, and usually requires a time-consuming dressing process. A small layer of the transmissive elastomer between the skin and the fiber tip has vastly increased the comfort and wear-time while maintaining skin contact and obtaining good signal levels.

Compared to reported wearable systems, which have been primarily used to examine prefrontal areas accessible in front of the hairline, this headgear will place nothing on the head but the interrogating fiber optics. This saves bulk for applications requiring lightweight, low-profile sensors (such as under a helmet or integrated with headphones), provides for the possibility of increased sensor population due to reduced footprint at each source and detector location, and allows compatibility for multimodal validation testing.

A second feature of the next-generation fNIRS headgear is a comb-based shape, which mounts the optical surfaces in the wake of a comb tooth so that the hair is automatically parted at each location that will be examined. Application is in a glasses-to-headband motion while maintaining contact between the optical surface and the skin, obviating the need for a second person to address the application and coupling of each optical sensor. Integration with existing fNIRS commercial instrumentation depends only on the fiber optic connection type. This aspect of the work is in the prototype phase.

The moldable properties of the elastomer provide the potential for an elegant solution employing diffractive optical technology. A grating embedded in the elastomer has the potential to turn the light toward the scalp over a very short distance without using a glass prism. This aspect of the work is in the design phase.

Outside of aeronautic and space applications, this research also facilitates investigations of neuroscience in practical work settings, and the development of usable brain-computer interfaces for biofeedback, rehabilitation, skill acquisition and self-treatment. The field of fNIRS is growing rapidly in both research and clinical applications and commercial fNIRS systems are emerging, with brain-computer interfaces becoming a reality in the market.

Science Mission Directorate

NASA leads the Nation on great journey of discovery, seeking new knowledge and understanding of our planet Earth, our Sun and solar system, and the universe out to its farthest reaches and back to its earliest moments of existence. NASA's Science Mission Directorate (SMD) and the Nation's science community use space observatories to conduct scientific studies of the Earth from space, to visit and return samples from other bodies in the solar system, and to peer out into our galaxy and beyond.

Kepler Finds Five New Exoplanets

NASA's Kepler space telescope, designed to find Earth-size planets in the habitable zone of Sun-like stars, has discovered its first five new exoplanets, or planets beyond our solar system. Kepler's high sensitivity to both small and large planets enabled the discovery of the exoplanets, named Kepler 4b, 5b, 6b, 7b, and 8b. Thediscoveries were announced January 4, 2010, by the members of the Kepler science team during a news briefing at the

NASA's Kepler space telescope has discovered five new planets beyond our solar system.

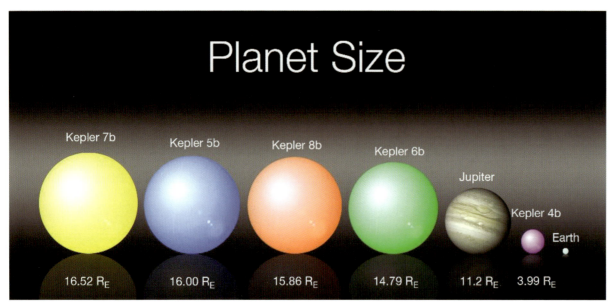

NASA's Kepler space telescope, designed to find Earth-size planets in the habitable zone of Sun-like stars, has discovered its first five new exoplanets beyond our solar system, all of which are large in relative size to the Earth (R_E).

American Astronomical Society meeting in Washington, D.C.

"These observations contribute to our understanding of how planetary systems form and evolve from the gas and dust disks that give rise to both the stars and their planets," said William Borucki of Ames Research Center. Borucki is the mission's science principal investigator. "The discoveries also show that our science instrument is working well. Indications are that Kepler will meet all its science goals."

Known as "hot Jupiters" because of their high masses and extreme temperatures, the new exoplanets range in size from similar to Neptune to larger than Jupiter. They have orbits ranging from 3.3 to 4.9 days. Estimated temperatures of the planets range from 2,200 to 3,000 °F, hotter than molten lava and much too hot for life as we know it. All five of the exoplanets orbit stars hotter and larger than Earth's Sun.

"It's gratifying to see the first Kepler discoveries rolling off the assembly line," said Jon Morse, director of the Astrophysics Division at NASA Headquarters. "We expected Jupiter-size planets in short orbits to be the first planets Kepler could detect. It's only a matter of time before more Kepler observations lead to smaller planets with longer period orbits, coming closer and closer to the discovery of the first Earth analog."

Launched on March 6, 2009, from Cape Canaveral Air Force Station in Florida, the Kepler mission continuously and simultaneously observes more than 150,000 stars. Kepler's science instrument, or photometer, already has measured hundreds of possible planet signatures that are being analyzed. While many of these signatures are likely to be something other than a planet, such as small stars orbiting larger stars, ground-based observatories have confirmed the existence of the five exoplanets.

The discoveries are based on approximately 6 weeks' worth of data collected since science operations began on May 12, 2009. Kepler looks for the signatures of planets by measuring dips in the brightness of stars. When planets cross in front of, or transit, their stars as seen from Earth, they periodically block the starlight. The size of the planet can be derived from the size of the dip. The temperature can be estimated from the characteristics of the star it orbits and the planet's orbital period.

Kepler will continue science operations until at least November 2012. It will search for planets as small as Earth, including those that orbit stars in a warm habitable zone where liquid water could exist on the surface of the planet. Since transits of planets in the habitable zone of solar-like stars occur about once a year and require three transits for verification, it is expected to take at least 3 years to locate and verify an Earth-size planet.

According to Borucki, Kepler's continuous and long-duration search should greatly improve scientists' ability to determine the distributions of planet size and orbital period in the future. "Today's discoveries are a significant contribution to that goal," Borucki said. "The Kepler observations will tell us whether there are many stars with planets that could harbor life, or whether we might be alone in our galaxy."

Kepler is the 10th mission of NASA's Discovery Program. Ames is responsible for the ground system development, mission operations, and science data analysis. NASA's Jet Propulsion Laboratory (JPL), managed the Kepler mission development. Ball Aerospace & Technologies Corporation, of Boulder, Colorado, was responsible for developing the Kepler flight system. Ball and the Laboratory for Atmospheric and Space Physics at the University of Colorado at Boulder, are supporting mission operations. Ground observations necessary to confirm the discoveries were conducted with these ground-based telescopes: Keck I in Hawaii; Hobby-Eberly and Harlan J. Smith in Texas; Hale and Shane in California; WIYN, MMT, and Tillinghast in Arizona; and Nordic Optical in the Canary Islands, Spain.

New Eye on the Sun Delivers Stunning First Images

NASA's recently launched Solar Dynamics Observatory, or SDO, is returning early images that confirm an unprecedented new capability for scientists to better understand our Sun's dynamic processes. These solar activities affect everything on Earth. Some of the images from the spacecraft show never-before-seen detail of material streaming outward and away from sunspots.

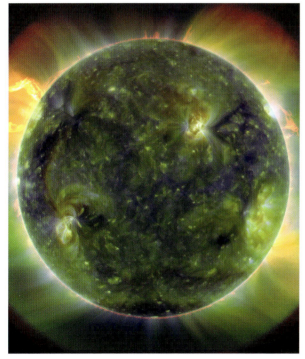

This full-disk multiwavelength extreme ultraviolet image of the Sun was taken by the Solar Dynamics Observatory (SDO) on March 30, 2010. False colors trace different gas temperatures. Reds are relatively cool (about 107,540 °F), while blues and greens are hotter (1,799,540 °F).

The Solar Dynamics Observatory is the most advanced spacecraft ever designed to study the Sun.

Others show extreme close-ups of activity on the Sun's surface. The spacecraft also has made the first high-resolution measurements of solar flares in a broad range of extreme ultraviolet wavelengths.

"These initial images show a dynamic Sun that I had never seen in more than 40 years of solar research," said Richard Fisher, director of the Heliophysics Division at NASA Headquarters. "SDO will change our understanding of the Sun and its processes, which affect our lives and society. This mission will have a huge impact on science, similar to the impact of the Hubble Space Telescope on modern astrophysics."

Launched on February 11, 2010, SDO is the most advanced spacecraft ever designed to study the Sun. During its 5-year mission, it will examine the Sun's magnetic field and also provide a better understanding of the role the Sun plays in Earth's atmospheric chemistry and climate. Since launch, engineers have been conducting testing and verification of the spacecraft's components. Now fully operational, SDO will provide images with clarity 10 times better than high-definition television and will return more comprehensive science data faster than any other solar observing spacecraft.

SDO will determine how the Sun's magnetic field is generated, structured, and converted into violent solar events such as turbulent solar wind, solar flares and coronal mass ejections. These immense clouds of material, when directed toward Earth, can cause large magnetic storms in our planet's magnetosphere and upper atmosphere.

SDO will provide critical data that will improve the ability to predict these space weather events. Goddard Space Flight Center built, operates, and manages the SDO spacecraft for SMD.

"I'm so proud of our brilliant work force at Goddard, which is rewriting science textbooks once again," said Senator Barbara Mikulski, D-Maryland, chairwoman of the Appropriations Subcommittee on Commerce, Justice, Science, and Related Agencies, which funds NASA. "This time Goddard is shedding new light on our closest star, the Sun, discovering new information about powerful solar flares that affect us here on Earth by damaging communication satellites and temporarily knocking out power grids. Better data means more accurate solar storm warnings."

Space weather has been recognized as a cause of technological problems since the invention of the telegraph in the 19th century. These events produce disturbances in electromagnetic fields on Earth that can induce extreme currents in wires, disrupting power lines and causing widespread blackouts. These solar storms can interfere with communications between ground controllers, satellites, and airplane pilots flying near Earth's poles. Radio noise from the storm also can disrupt cell phone service.

SDO will send 1.5 terabytes of data back to Earth each day, which is equivalent to a daily download of half a million songs onto an MP3 player. The observatory carries three state-of-the-art instruments for conducting solar research.

The Helioseismic and Magnetic Imager (HMI) maps solar magnetic fields and looks beneath the Sun's opaque surface. The experiment will decipher the physics of the Sun's activity, taking pictures in several very narrow bands of visible light. Scientists will be able to make ultrasound images of the Sun and study active regions in a way similar to watching sand shift in a desert dune. The instrument's principal investigator is Phil Scherrer of California's Stanford University. HMI was built through a collaboration of Stanford University and the Lockheed Martin Solar and Astrophysics Laboratory (LMSAL) in Palo Alto, California.

The High Resolution Imaging Science Experiment camera on NASA's Mars Reconnaissance Orbiter took these images of a fresh, 20-foot-wide crater on Mars on October 18, 2008 (left) and on January 14, 2009 (right). Each image shows an area 115 feet across, and the crater's depth is estimated to be 4.4 feet.

The Atmospheric Imaging Assembly is a group of four telescopes designed to photograph the Sun's surface and atmosphere. The instrument covers 10 different wavelength bands, or colors, selected to reveal key aspects of solar activity. These types of images will show details never seen before. The principal investigator is Alan Title of the LMSAL, which built the instrument.

The Extreme Ultraviolet Variability Experiment measures fluctuations in the Sun's radiant emissions. These emissions have a direct and powerful effect on Earth's upper atmosphere—heating it, puffing it up, and breaking apart atoms and molecules. Researchers don't know how fast the Sun can vary at many of these wavelengths, so they expect to make discoveries about flare events. The principal investigator is Tom Woods of the Laboratory for Atmospheric and Space Physics, at the University of Colorado at Boulder, which built the instrument.

"These amazing images, which show our dynamic Sun in a new level of detail, are only the beginning of SDO's contribution to our understanding of the Sun," said SDO project scientist Dean Pesnell of Goddard.

SDO is the first mission of NASA's Living with a Star Program, or LWS, and the crown jewel in a fleet of NASA missions that study our Sun and space environment. The goal of LWS is to develop the scientific understanding necessary to address those aspects of the connected Sun-Earth system that directly affect our lives and society.

Underground Ice on Mars Exposed by Impact Crater

Images taken by the Thermal Emission Imaging System camera on NASA's Mars Odyssey orbiter and by the Context Camera on the Mars Reconnaissance Orbiter show exposed water ice from below the surface created by a recent impact that excavated a crater. The water ice is seen as bright material around the crater's edge in an image taken in October 2008, and then nearly gone in the following image taken in January 2009. The change in appearance resulted from some of the ice sublimating away during the Martian northern-hemisphere summer, leaving behind dust that had been intermixed with the ice. The thickening layer of dust on top obscured the remaining ice.

Analysis of the observations of fresh craters exposing ice reported by Byrne et al. in a September 25, 2009, paper in the journal *Science*, leads the paper's authors to calculate that if NASA's Viking Lander 2 had been able to dig slightly deeper than the 4- to 6-inch-deep trench that it excavated in 1976, it would have hit water ice. Discovery of water ice in large quantities close to the surface and near the equator of Mars (where landing is easier), may make a profound difference for any future human exploration missions to the Red Planet.

Earth Sciences Division Responds to Several Emergencies in 2010

The integrated capability of the space and air systems operated by the Earth Sciences Division had a great year of science, but also responded to many unique challenges to the people and environment of the planet. In most of these cases, many systems were brought into play as a team effort, including the oil spill in the Gulf, the volcano eruption in Iceland, and the earthquake in Haiti.

At the request of U.S. disaster response agencies, an advanced JPL-built optical sensor flying aboard a NASA research aircraft was among several NASA remote-sensing assets mobilized to help assess the spread and impact of the Deepwater Horizon BP oil spill in the Gulf of Mexico.

As part of the national response to the spill, and at the request of the National Oceanic and Atmospheric Administration (NOAA) and the U.S. Geological Survey (USGS), NASA deployed an instrumented research air-

NASA's Earth Resource-2 research aircraft, with the Jet Propulsion Laboratory's advanced Airborne Visible/Infrared Imaging Spectrometer instrument aboard, flew from California to Texas on May 6, 2010, for a series of flights to map the Gulf of Mexico oil spill and coastal areas.

craft, the Earth Resources-2 (ER-2), to the Gulf on May 6, 2010. The ER-2, outfitted with JPL's Airborne Visible/Infrared Imaging Spectrometer (AVIRIS) and the Cirrus Digital Camera System, supplied by Ames, was sent to collect detailed images of the Gulf of Mexico and its threatened coastal wetlands.

NASA also made extra satellite observations and conducted additional data processing to assist NOAA, the U.S. Geological Survey (USGS), and the U.S. Department of Homeland Security in monitoring the spill.

"NASA has been asked to help with the first response to the spill, providing imagery and data that can detect the presence, extent, and concentration of oil," said Michael Goodman, program manager for natural disasters in the Earth Science Division of NASA's SMD. "We also have longer-term work we have started in the basic research of oil in the ocean and its impacts on sensitive coastal ecosystems."

NASA pilots flew the ER-2 from Dryden Flight Research Center in California to a temporary base of operations at Johnson Space Center's Ellington Field in Houston. Along the way, the plane collected data over the Gulf coast and the oil slick to support spill mapping and to document the condition of coastal wetlands before oil landfall. The ER-2 made a second flight on May 10, and combined NASA aircraft flew 24 missions and acquired more than 120 hours of data over the course of 4 deployments from May through July.

The AVIRIS team, led by JPL's Robert Green, measured how the water absorbs and reflects light in order to map the location and concentration of oil, which separates into a thin, widespread sheen and smaller, thick patches. Satellites can document the overall extent of the oil but cannot distinguish between the sheen and thick patches. While the sheen represents most of the area of the slick, the majority of the oil is concentrated in the thicker part. AVIRIS has the capability to identify the thicker parts, helping oil spill responders know where to deploy oil-skimming boats and absorbent booms. Researchers also planned to measure changes in vegetation along the coastline and assess if, where, and how oil affected marshes, swamps, bayous, and beaches that are difficult to survey on the ground. The combination of satellite and airborne imagery assisted NOAA in forecasting the trajectory of the oil and in documenting changes in the ecosystem.

From the outset of the spill on April 20, 2010, NASA provided satellite images to Federal agencies from the Moderate Resolution Imaging Spectroradiometer (MODIS) instruments on NASA's Terra and Aqua satellites; the Japanese Advanced Spaceborne Thermal Emission and Reflection Radiometer (ASTER) on Terra; and the Advanced Land Imager (ALI) and Hyperion instruments on NASA's Earth Observing-1 (EO-1) satel-

Ash and steam billowed from the Eyjafjallajökull volcano in early May 2010. The Advanced Land Imager on NASA's Earth Observing-1 satellite captured this natural-color image on May 2, 2010.

lite. All of these observations were funneled to the Hazards Data Distribution System operated by the USGS. With its very wide field of view, MODIS provided a big picture of the oil spill and its evolution roughly twice a day. The Hyperion, ALI, and ASTER instruments observed over much smaller areas in finer detail, but less often (every 2 to 5 days).

Other NASA satellite and airborne instruments collected observations of the spill to advance basic research and to explore future remote-sensing capabilities. From space, the JPL-built and managed Multi-angle Imaging Spectroradiometer instrument on Terra, JPL's Atmospheric Infrared Sounder instrument on Aqua, and the Cloud-Aerosol Lidar with Orthogonal Polarization (CALIOP) on the joint NASA-France CALIPSO satellite collected data. Another NASA research aircraft, the

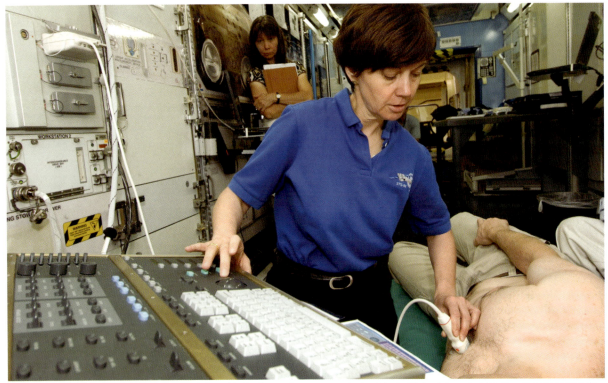

Astronaut Cady Coleman performs a remotely guided echocardiogram on a test subject utilizing Integrated Cardiovascular protocols while Betty Chen, a training coordinator, observes.

In response to the disaster in Haiti on January 12, 2010, NASA added a series of flights over earthquake faults in Haiti and the Dominican Republic on the island of Hispaniola. NASA's Uninhabited Aerial Vehicle Synthetic Aperture Radar, or UAVSAR, left Dryden on January 25, 2010, aboard a modified NASA Gulfstream III aircraft. During its trek to Central America, which ran through mid-February, the repeat pass L-band wavelength radar, developed by JPL, studied the structure of tropical forests; monitored volcanic deformation and volcano processes; and examined Mayan archaeology sites.

After the Haitian earthquake, NASA managers added additional science objectives that will allow UAVSAR's unique observational capabilities to study geologic processes in Hispaniola. UAVSAR's ability to provide rapid access to regions of interest, short repeat flight intervals, high resolution, and its variable viewing geometry make it a powerful tool for studying ongoing Earth processes.

Space Operations Mission Directorate

The Space Operations Mission Directorate (SOMD) provides NASA with leadership and management of the Agency's space operations related to human exploration in and beyond low Earth orbit. Space Operations also oversees low level requirements development, policy, and programmatic oversight. Current exploration activities in low Earth orbit include the space shuttle and International Space Station (ISS) programs. The directorate is similarly responsible for Agency leadership and management of NASA space operations related to launch services, space transportation, and space communications in support of both human and robotic exploration programs. Its main challenges include: completing assembly of the ISS; utilizing, operating, and sustaining the ISS; commercial space launch acquisition; future space communications architecture; and transition from the space shuttle to future launch vehicles.

King Air B200 from Langley Research Center also collected data over the area of the oil spill. It completed its first flight over the spill on May 10. The High Spectral Resolution Lidar onboard the plane uses pulses of laser light to locate and identify particles in the environment. Led by Chris Hostetler of Langley, the lidar provides measurements similar to those from the CALIOP instrument. Data from these space-based and airborne lidars were used to investigate the thickness of the oil spill below the surface of the water and evaluate the impacts of dispersants used to break up the oil.

"Although NASA's primary expertise is in using remote-sensing instruments to conduct basic research on the entire Earth system, our observations can be used for societal benefit in response to natural and technological disasters like this oil spill," said Goodman.

On May 4, 2010, the Icelandic Meteorological Office warned that Eyjafjallajökull showed no signs of ending its eruptive activity in the near future. The office reported that ash from the volcano had reached a height of 19,000 to 20,000 feet above sea level, and had spread 40 to 50 miles east-southeast of the volcano, where it impeded visibility for local residents. The office also reported that lava continued flowing down a steep hill north of the crater. The ALI on NASA's EO-1 satellite captured images of the volcano.

Research on Astronauts' Muscles Gets to the Heart of the Matter

When humans venture into space for long periods, muscles tend to weaken because they do not have to work as hard without gravity. Of course, the most important muscle is the heart.

While doctors are well aware of the weakening of the heart in space, known as cardiac atrophy, a new study on the ISS seeks to find out exactly how much of the heart muscle decreases in size over a standard 6-month station tour, and how quickly it occurs.

In addition to evaluating cardiac health in space, the Integrated Cardiovascular investigation also will determine how effective the astronauts' current exercise program is at protecting the heart from getting smaller or weaker.

"This study also will help us determine if there is a risk of abnormal heart rhythms and how significant the risk is in order to develop appropriate countermeasures," said Dr. Deborah Harm, the international project scientist for the ISS Medical Program at Johnson Space Center.

According to Harm, many crewmembers experience a brief period of lightheadedness and a drop in blood pressure when standing still after coming home to Earth from long-duration missions. Fainting can occur when the heart cannot generate enough force to pump the necessary blood to the brain and the rest of body—either because the muscle is too small or weak, or because there is an abnormal heart rhythm.

"At this time, it is unknown if heart muscle weakening continues throughout a mission or if it levels off at some point. That's what we want to find out," Harm said.

Crewmembers on Expedition 20 (2009) were the first to participate. Before, during, and after flight, they measured their heart rates, heart rhythms, and blood pressure for 24 to 48 hours before and after exercise sessions. They also performed on-orbit cardiac ultrasound scans on each other before and after exercise to look at how effectively

NASA astronaut Nicholas Patrick, STS-130 mission specialist, participates in the mission's third and final session of extravehicular activity (EVA) as construction and maintenance continue on the ISS. During the 5-hour, 48-minute spacewalk, Patrick and astronaut Robert Behnken (out of frame), mission specialist, completed all of their planned tasks, removing insulation blankets and removing launch restraint bolts from each of the Cupola's seven windows.

the heart fills with blood and pumps it to the rest of the body.

"MRI scans were done on crewmembers' hearts before and after flight to measure exactly how much heart muscle was present and will be compared to the cardiac ultrasound information to better understand how changes in heart muscle are related to cardiac function," said Dr. Michael Bungo of the investigator team.

"Such an extensive and sophisticated study of the cardiovascular system was virtually impossible before we had six crewmembers onboard the station," Harm added. "There simply was not enough crew time available to complete all the procedures required for this experiment."

While in space, crewmembers wore four devices: a portable Holter monitor that measures heart rate continuously for extended periods; a Cardiopres that measures blood pressure with every heart beat; and two Actiwatches—one on an ankle and one on a wrist—to monitor and record body movements.

The data collected was transmitted to the Payload Operations Center at Marshall Space Flight Center, and delivered to the investigator team for analysis.

This study shows the breadth of international cooperation and collaboration that occurs on the space station. Three international partner agencies are working together to get the best science. The European Space Agency (ESA) provided the Cardiopres device for monitoring blood pressure, and the investigators shared the Holter data with teams for two Canadian-sponsored experiments. One of these experiments also includes ESA investigators.

All of these teams are studying different aspects of the cardiovascular system. Sharing this data among scientists greatly enhances the overall science return. "This allows us to more efficiently and quickly understand the full range of cardiovascular changes than any one investigation could," Harm said.

Knowledge gained in the Integrated Cardiovascular study may help doctors treat patients on Earth who have been confined or on bed rest for long periods. Patients with heart diseases that change their normal cardiac function may also benefit.

New Experiment Aboard ISS Smiles on the Environment

There's a new way to look at environmental issues on Earth—from 210 miles up onboard the ISS—and investigators are smiling about the results. The SMILES experiment, more properly known as the Superconducting Submillimeter-wave Limb-emission Sounder, is investigating issues such as ozone depletion and air quality problems.

The experiment launched on the Japan Aerospace Exploration Agency's (JAXA) H-II Transfer Vehicle—an unmanned cargo ship for station resupply. Housed on the Japanese Experiment Module's Exposed Facility, SMILES is gathering data on trace gasses known to cause ozone depletion, such as chlorine and bromine compounds. The facility provides a multipurpose platform where science experiments can be deployed and operated in open space. The observations are taken in the stratosphere, the region of the atmosphere 6 to 30 miles above the Earth's surface.

"Measurements of ozone and trace gasses in the stratosphere from instruments such as SMILES are important for understanding the dynamics of Earth's atmosphere," said Julie Robinson, ISS program scientist at Johnson.

The fruit of *Jatropha curcas* plants contain three seeds that can be pressed for oil to be used as biofuel. This Fluid Processing Apparatus (top right) contains cell suspensions of *Jatropha curcas*. It was assembled into the Group Activation Pack, which was transported to the ISS for microgravity studies.

The advantage of this experiment is the space station's power and payload resources, which enable researchers to test out new technologies. As a result, SMILES can measure precise molecules of trace atmospheric gasses and obtain data about elements in quantities too small to be measured until now.

SMILES observations taken in October 2009 show that ozone amounts are greater around Earth's equatorial region than at higher latitudes, illustrating the characteristics of stratospheric ozone in its global distribution.

"This is just the beginning," said Takuki Sano, a member of the SMILES science team with JAXA. "In due course, SMILES, with its full-scale observation, will contribute to the prediction of ozone depletion through analyses of the accumulated observation data, thus clarifying the influence the stratosphere has on the troposphere—the lowest and most dense layer of the atmosphere 10 to 12 miles above the Earth's surface."

Alternative Energy Crops Grow in Space

What if space held the key to producing alternative energy crops on Earth? That's what researchers are hoping to find in a new experiment on the ISS.

The experiment, National Lab Pathfinder-Cells 3, is aimed at learning whether microgravity can help *Jatropha curcas* plant cells grow faster to produce biofuel, or renewable fuel derived from biological matter. *Jatropha* is known to produce high-quality oil that can be converted into an alternative energy fuel, or biofuel.

By studying the effects of microgravity on *Jatropha* cells, researchers hope to accelerate the cultivation of the plant for commercial use by improving characteristics such as cell structure, growth, and development. This is the first study to assess the effects of microgravity on cells of a biofuel plant.

"As the search for alternate energy sources has become a top priority, the results from this study could add value for commercialization of a new product," said Wagner Vendrame, principal investigator for the experiment at the University of Florida in Homestead. "Our goal is to verify if microgravity will induce any significant changes in the cells that could affect plant growth and development back on Earth."

Launched on space shuttle Endeavour's STS-130 mission in February 2010, cell cultures of *Jatropha* were sent to the space station in special flasks containing nutrients and vitamins. The cells were exposed to microgravity until they returned to Earth aboard space shuttle Discovery's STS-131 mission.

For comparison studies of how fast the cultures grow, a replicated set of samples are being maintained at the University of Florida's Tropical Research and Education Center in Homestead.

"Watching the space shuttle go up carrying a little piece of my work is an indescribable experience," said Vendrame. "Knowing that my experiment could contribute to creating a sustainable means for biofuel production on Earth, and therefore making this a better world, adds special value to the work."

ISS Expansion Includes a Room with a View

For the past several years, the ISS has been moving steadily closer to completion. But what house is complete without a utility room, a gym, and a picture window?

During the STS-130 mission, Space Shuttle Endeavour delivered the Tranquility node and its cupola, a dome-shaped extension from Tranquility made up of seven windows. They are the last major U.S. modules to be added to the space station, and together they helped clear out premium workspace in other areas of the station—as well as offer a window on the world.

At 15-feet wide and 23-feet long, the Tranquility node provides a centralized home for the station's environmental control equipment—one of the systems that remove carbon dioxide from the station's air, one of the station's bathrooms and the equipment that converts urine into drinkable water, all of which previously took up space in the Destiny laboratory. And there's enough room left over to house the station's new treadmill and its microgravity equivalent of a weight machine, moving it out of the Unity node where it was in the way whenever spacewalk preparations were going on inside the adjacent Quest airlock.

"Out the window is the truth."

"It [gave] us a much needed addition to the house, so to speak," said Bob Dempsey, lead space station flight director for the mission. "[We were] getting to the point where we're really cramped for space. You might be surprised at that, considering we're essentially the volume of a 747 and we've been adding modules for the last couple

NASA astronauts Terry Virts (left), STS-130 pilot; and Jeffrey Williams, Expedition 22 commander, pose for a photo near the windows in the newly-installed Cupola of the ISS while Space Shuttle Endeavour remains docked with the station.

NASA astronaut Nicholas Patrick, STS-130 mission specialist, is pictured in a window of the newly-installed Cupola of the ISS.

of years. You might think we'd be sitting around in a big empty house. But no—every inch is really getting packed up there."

STS-130 Commander George Zamka put it another way. "It's like exercising in the office," he said. "This will be a more logical organization, more focused."

Though the node has an intensely practical function, there are still fanciful aspects to Tranquility, including its name, which was chosen with the help of a naming contest on www.nasa.gov.

"It harkens back to the Sea of Tranquility, where humans made their very first tentative landing on the Moon," Zamka said. "They were only there for a few hours, and it was at the very limits of what human beings could do. From that beginning, we're now putting up a node that will house the majority of the life support equipment for the station, where we're going to have a permanent presence in space."

But everyone agrees that the real scope for the imagination will be provided by Tranquility's 6.5- by 5-foot annex: the cupola. Its purpose is to provide a true view of robotics operations on the station's exterior—such as those that will be required when the next module, the Russian Rassvet, is added during STS-132.

"Out the window is the truth," Zamka said. "The video views that we use now, you're trying to stick together and have a mental image of where things are. When you look out the window, you don't have to imagine. It's all right there for you."

But there's no question that many people, including Zamka, are looking forward to looking out of it for other views.

"Just the idea of providing this great view of the station and the world beneath us is going to be pretty great," he said. "That's not what it's for, but it will be spectacular."

The cupola will be like a mini control tower sticking out from the Tranquility node, as opposed to the other station windows, which are flush with the station's exterior. Its seven windows—one in the center and six around the sides—will provide the only views of the outside of the station from the inside, in particular the Russian and Japanese sections. And with the station just about finished, there's more to see out there than ever.

So, Zamka said, in addition to the robotic operations and Earth views it will provide, it will also give us a good look at some of the space shuttle fleet's finest handiwork as the program comes to an end. And that provides its own cause for reflection.

"We've come a long way in human space flight because of the shuttle's capability," he said. "We've launched and retrieved satellites, we've done medical research, and now we've built this huge space station. We're almost to the point of passing the baton from the space shuttle to the space station in terms of what our human space flight experience will be now."

Kwatsi Alibaruho, the lead STS-130 space shuttle flight director, said that even with so much left to do in the program's final five flights, he was making it a point to spend some time thinking about the subject.

"It's very easy to get into a routine, to lose oneself in the hustle and bustle of trying to get the work done," Alibaruho said. "But the shuttle is a unique spacecraft. I find myself thinking a lot about how I'm going to describe this time to my son when he's old enough to understand. There has never been an operational spacecraft like it before and all indications are that it will be some time before there will be one like it again. I find myself really appreciative of the opportunity I've had to serve in this capacity."

Boeing Delivers 'Keys' to ISS

In March of 2010, NASA officially accepted the "keys" to the ISS from its prime contractor, Boeing, at the conclusion of an Acceptance Review Board (ARB) that verified the delivery, assembly, integration, and activation of all hardware and software required by contract.

"The successful completion of this ISS contract is a testament to the hard work, dedication, and perseverance of an amazing international team of government agencies and their commercial contractors," said ISS program manager Michael Suffredini.

"I want to congratulate the entire Boeing team, including its many suppliers and subcontractors, for their service to NASA and the world," Suffredini added. "As we near completion of this orbiting laboratory, we are only beginning to understand its true value as the dividends in our investment pay off with advances in medicine, technology, and international relations."

The ARB was an administrative formality that culminated in submission of government form DD 250, in which Boeing confirmed, and NASA accepted, that all major contract requirements have been met. In effect, the DD 250 transfers station ownership to NASA. The ARB examined in exhaustive detail the past and current performance since the first element was launched in 1998.

The review came on the heels of the STS-130 mission of Endeavour, which delivered the Tranquility module and cupola, the final living areas of the U.S. On-orbit Segment (USOS). The USOS incorporates all contributions to the station by NASA, ESA, the Canadian Space Agency, the JAXA, and interfaces with the Russian On-Orbit Segment, which includes the components provided by the fifth partner, the Russian Federal Space Agency.

The Sahara Desert spread out through the array of windows. The Cupola will house controls for the station robotics and will be a location where crew members can operate the robotic arms and monitor other exterior activities.

The football field-sized outpost is now 90-percent complete by mass, and 98-percent complete by internal volume. Supporting a multicultural crew of six, the station has a mass of almost 400 tons and more than 12,000 cubic feet of living space.

Upon completion of assembly, the station's crew and its U.S., European, Japanese, and Russian laboratory facilities will expand the pace of space-based research to unprecedented levels. Nearly 150 experiments are currently underway on the station, and more than 400 experiments have been conducted since research began 9 years ago. These experiments already are leading to advances in the fight against food poisoning, new methods for delivering medicine to cancer cells, and the development of more capable engines and materials for use on Earth and in space. ❖

Access® and Excel® are registered trademarks of Microsoft Corporation.

Kevlar® is a registered trademark of E. I. du Pont de Nemours and Company.

Education News

Using the inspirational pursuit of space exploration to spark the imagination of our youth is critical to creating a scientifically literate populace and keeping this Nation competitive in an ever more technologically advanced world. Toward that end, NASA has a tradition of investing in programs and activities that engage students, educators, families, and communities in exploration and discovery—helping support the Agency's missions and cultivate its future leaders and innovators.

Education News

NASA Supports the President's 'Educate to Innovate' Campaign with Summer of Innovation

In January 2010, NASA launched an initiative to use its out-of-this-world missions and technology programs to boost summer learning, particularly for underrepresented and underperforming students across the Nation. NASA's Summer of Innovation supports President Obama's "Educate to Innovate" campaign for excellence in science, technology, engineering, and mathematics (STEM) education.

The Summer of Innovation program engaged thousands of middle school teachers and students in stimulating math- and science-based education programs.

In April, NASA announced that it is partnering with the Space Grant Consortia of Idaho, Massachusetts, New Mexico, and Wyoming in the Summer of Innovation initiative. NASA awarded four cooperative agreements that total approximately $5.6 million. Awards have a period of performance of 36 months. Local programs are required to develop ways to keep students and teachers engaged during the school year and to track student participants' performance through 2012. Awardees are encouraged to leverage the unique capabilities and resources of program partners to ensure a sustainable effort following the period of performance.

Kicked off in early June, the Summer of Innovation program engaged thousands of middle school teachers and students in stimulating math- and science-based education programs. NASA's goal is to increase the number of future scientists, mathematicians, and engineers, with an emphasis on broadening participation of low-income and minority students.

In response to President Obama's "Educate to Innovate" campaign to improve the participation and performance of America's students in science, technology, engineering, and mathematics (STEM), NASA launched its Summer of Innovation program.

The Summer of Innovation was conducted through a multifaceted approach that allowed NASA to assess the viability, scalability, and success of the pilot programs. Now that the Summer of Innovation pilot has concluded, NASA is conducting an analysis to determine the best practices to build capacity to implement a comprehensive project in the summer of 2011 and beyond.

NASA used the Agency's National Space Grant College and Fellowship Program to implement the Summer of Innovation pilot. The Space Grant national network consists of 52 consortia in all 50 states, the District of Columbia, and the Commonwealth of Puerto Rico. The network includes 850 affiliates from universities, colleges, industry, museums, science centers, and state and local agencies supporting and enhancing science and engineering education, research, and public outreach efforts for NASA's aeronautics and space projects.

Educator Astronaut Operates Robotic Arm in Space

Dottie Metcalf-Lindenburger launched to space on April 5, 2010 aboard the STS-131 mission for her first space flight but will be the last of three school teachers to fly on the space shuttle as mission specialists from the 2004 Educator Astronaut Class. The first two, Ricky Arnold and Joseph Acaba, flew on the STS-119 shuttle mission in March 2009.

The educational activities on the STS-131 shuttle mission to the International Space Station (ISS) focused on robotics and careers in science, technology, engineering, and math.

Without robotics, major accomplishments of building the station, repairing satellites in space, and exploring other worlds would not be possible. While in space, Metcalf-Lindenburger operated the space shuttle's robotic arm and a 50-foot Orbiter Boom Sensing System to inspect the shuttle for any damage that might have occurred during launch or in space. A digital camera and laser system on the boom's end provided three-dimensional imagery used by analysts to assess the health of the shuttle's heat shield.

Already, robotic arms have made it possible for NASA to accomplish amazing feats, and Metcalf-Lindenburger predicts robotic devices will continue to be developed and used for future space missions. Robots may be used to collect and analyze samples or help with the construction of an outpost.

Today's students will be the ones building the robots of the future, and some students have already started. Children as young as age 6 are participating in student robotics competitions. As one of NASA's educator astronauts, Metcalf-Lindenburger believes it is important for students to understand the potential importance of robotics.

Robotics will be important not only to the future of space exploration but in all types of fields, Metcalf-Lindenburger said. "You see it being used in auto manufacturing and in medicine, so I think we'll continue to see robotics used more and more in society. Students need to be aware of how robotics can be used for different things."

For educators, Metcalf-Lindenburger said robotics can be a fun way to engage students in STEM subjects. "This is a way to say, 'This is important to study math and science, because you may be working with these things, and you may be designing these things. You will probably be influenced by robots at some time in your life.'"

Office of Education Supports NASA Open Government Movement

In his first executive action, President Obama released the Open Government Memorandum, which called for a more transparent, participatory, and collaborative government. NASA has successfully responded to the directive with comprehensive action, and its Office of Education actively participated in this Agency-wide effort. Open Government is of particular importance to this functional office, as the Agency's founding legislation mandates that NASA should disseminate information to the public about its missions and discoveries.

The Agency recognizes the need to interact with these important stakeholders, and NASA's goals in education reflect this commitment to engaging the public. NASA education Web sites facilitate online participation and collaboration, and a wide variety of sites support participatory exploration, access to information, engagement, social media tools, and submission of feedback. Online mechanisms for internal and external reporting of NASA education activities further promote teamwork, transparency, collaboration, and awareness. For example, the Office of Education uses an electronic Weekly Activity Report (WAR) tool to capture and report significant activities and upcoming events that are taking place across the Agency. It facilitates internal transparency and promotes efficiency and effectiveness by reducing redundancies between groups working towards similar goals from geographically dispersed locations. The Office of Education recently conducted activities to expand availability of this resource within the Agency.

The NASA Office of Education supported the Agency Open Government Working Group that was established in response to the White House directive. The Office of Education created a page on its Web site dedicated to Open Government and feedback from the public. In March 2010, an Office of Education Open Government fact sheet outlining current and future activities was developed for the Agency Open Government Plan

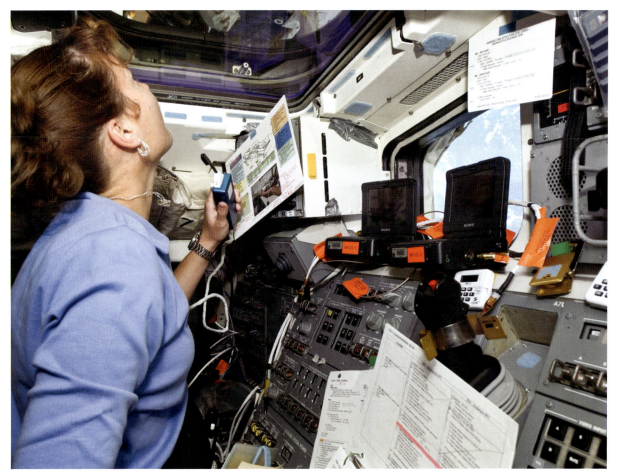

Educator astronaut Dottie Metcalf-Lindenburger, a fully qualified astronaut with expertise as a classroom teacher, operated the robotic arm aboard the space shuttle during STS-131.

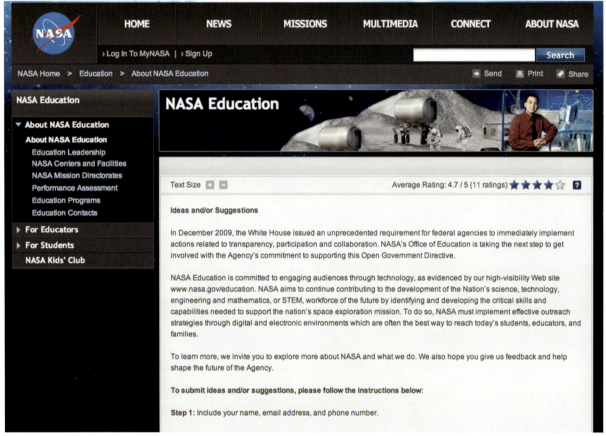

In support of President Obama's Open Government Initiative, NASA's Office of Education makes available most of its activities through the Internet, with added engagement through interactive feedback functions and social media tools.

released April 7, 2010. Education staff also developed and disseminated eight education specific questions for public feedback, and moderated the submissions during February and March. Consequently, 18,000 educators received an e-mail message with the education questions and 1.3 million Twitter followers received a link to the online feedback tool. As a result of these efforts, a significant amount of feedback related to education was received by NASA. The Agency received 453 ideas from 280 different authors through the online Citizen Engagement Tool. Of those, 126, or 28 percent, were things that NASA could legally and feasibly address in either short- or long-term plans. The Office of Education received 47 viable ideas after detailed analysis of submissions.

Immediate actions taken as of May 2010 in support of goals outlined in the education section of the Open Government Plan include the establishment of online collaborative tools as part of education events and activities, and preliminary submission of the Education WAR tool to the data.gov catalog. Near-term plans included a comprehensive internal and external communications plan based on Open Government values, and development of strategies to provide feedback to the public regarding the results of the idea submissions and other collaborations.

NASA Helps National Lab Day Lift Off

NASA demonstrated its commitment to STEM education and hands-on learning through activities supporting the first National Lab Day, a grassroots initiative aimed at bringing STEM hands-on activities to students across the country.

To kick off the week of the first annual National Lab Day on May 12, several cabinet and senior administration officials joined in to promote the "Educate to Innovate" campaign for excellence in STEM. NASA education staff supported two of these high-profile National Lab Day events in the District of Columbia.

- On Tuesday, May 11, Charles Bolden, NASA administrator, spoke about working in space to fifth graders who were studying the solar system at Langdon Education Campus. He had the opportunity to share his experience of living and working in space as a NASA astronaut and then becoming the NASA administrator. He had an energetic discussion with students on how Newton's laws are present in everyday life and also participated in a demonstration of a hands-on rocketry experiment.

- On Wednesday, May 12, Dr. John Holdren, director of the Office of Science and Technology Policy, worked with 40 physics students at Benjamin Banneker High School. After describing what it's like to be President Obama's science and technology advisor, Holdren took several engaging questions from the student audience. Following the Q&A, Holdren joined the students in an educational activity set up by NASA. Students created a simulated asteroid surface

NASA Administrator, Charles Bolden, participated in National Lab Day by speaking to a group of fifth graders at the Langdon Education Campus in Washington D.C. This was just one of the ways that NASA participated in this national event.

using a mixture of soil, flour, and other ingredients. Then, using golf balls and a protractor, they observed how changes in the angle of a projectile's impact affected the area and volume of the resulting craters.

- On Wednesday, May 12, Arne Duncan, secretary of U.S. Department of Education, joined third graders at the Martin Luther King Jr. Elementary School for an interactive session about science and engineering in their classroom. Students presented their designs of "shock absorbers" that any spacecraft would need for landing on the Moon or Mars. The engineering activity led to a discussion about energy and friction as students assembled and tested model cars on various surfaces.

Although many activities took place on National Lab Day, the event encompasses more than just 1 day. It is a nationwide collaborative movement that gets volunteers, university students, scientists, and engineers to work together with educators. Prior to its support of the events, NASA also hosted a series of weekly live webcasts during the month of April through the Agency's Digital Learning Network. The series aimed to equip teachers to promote hands-on science education in their classrooms. NASA's Digital Learning Network allows the next generation of explorers to connect with scientists, engineers, and researchers without leaving the classroom. Through interactive videoconferencing available at all 10 NASA field centers, the network provides distance-learning events designed to educate through demonstrations and real-time interactions with NASA experts.

NASA embraced National Lab Day and scheduled activities at schools throughout the week supported by volunteers from its field centers across the Nation and from its headquarters in Washington D.C. For instance, Kennedy Space Center hosted an educational event for students from local-area high schools who learned about NASA and the benefits of STEM fields in our world and beyond.

Bolden remarked positively about National Lab Day in an entry on the Office of Science and Technology Policy blog after the Langdon Education Campus event on May 11. He said, "There is a crisis in the United States that stems from the gap between the Nation's growing need for scientists, engineers, and other technically skilled workers, and our supply. This crisis in education, if not resolved, will contribute to future declines in qualified employees to meet demands in critical career fields that affect U.S. global competitiveness and the national economy. However, seeing the engagement and enthusiasm of those fifth grade students, I am hopeful that given the opportunity, our youth shall be inspired and motivated to consider STEM careers."

> *"There is a crisis in the United States that stems from the gap between the Nation's growing need for scientists, engineers, and other technically skilled workers, and our supply."*

NASA Funds Initiative to Develop One-Stop Internship Shopping

Undergraduate and graduate students who want to apply for a NASA internship or fellowship soon will have access to all of NASA's opportunities at one Web site. In April 2010, NASA announced that it had awarded cooperative agreements to five organizations to help attract students to NASA opportunities and provide an easily navigable process from start to finish.

The awards will fund a NASA-wide recruitment, application, selection, placement, and career development system to engage students in STEM research, aerospace education, and space exploration.

The objective of the One Stop Shopping Initiative is to provide an Agency-wide integrated system. The goal is a central destination for NASA science and engineering mentors to announce their internship and fellowship opportunities. The effort will consolidate announcements and information currently found on multiple NASA Web sites. The initiative also will formalize the transition of student participants in NASA programs into the workforce, including within NASA, the aerospace industry, and academia.

The Ohio Aerospace Institute in Cleveland will partner with NASA for business management. NASA selected four additional organizations to implement recruitment, retention, and career development strategies that broaden the diversity of institutions and individuals who apply for NASA's internships and fellowships. The organizations chosen to comprise the Broker Facilitator Corps are the Institute for Broadening Participation of Damariscotta, Maine; the United Negro College Fund Special Program of Falls Church, Virginia; the Hispanic College Fund Inc. of Washington, D.C.; and the American Indian Higher Education Consortium of Alexandria, Virginia. The total value of all awards is approximately $9.7 million during a 5-year funding period.

This program continues the Agency's efforts to develop highly qualified undergraduate and graduate students who possess skills in the STEM disciplines critical to creating a high technology workforce for NASA and the Nation.

Stellar Students Selected as NASA Ambassadors

In January 2010, NASA announced the selection of 105 of its best and brightest interns and fellows for the NASA Student Ambassador Program. The Agency uses the program to engage undergraduate and graduate students in NASA STEM research and interactive opportunities. Selected students represent 33 states and 81 universities from across the Nation.

NASA managers and mentors nominated the recipients from the hundreds of interns and fellows engaged in research and education opportunities across the Agency. The NASA Student Ambassadors initiative further recognizes exceptional students. Members of the NASA Student Ambassadors virtual community will interact with the Agency while sharing information, making professional connections, and collaborating with peers. They also will represent NASA in a variety of venues and help the Agency inspire and engage future interns and fellows.

The community's Web site provides participants with access to the tools needed to serve as a NASA Student Ambassador. The site provides strategic communication opportunities, the latest NASA news, science and technology updates, blogs, and announcements. It contains member profiles, forums, polls, NASA contact information, links to Agency mission-related communications' research, and career resources.

"The virtual community Web site is an outreach vehicle to the Nation's students as well as a way to engage exceptional Gen-Y NASA students," said Mabel Matthews, lead for the community and manager of higher education at NASA Headquarters. "This activity is a leading effort to help NASA attract, engage, educate, and employ the next generation."

Undergraduate and graduate students who enroll in NASA's internship and fellowship programs gain hands-on experience working on real NASA projects.

With this and the Agency's other college and university programs, NASA will identify and develop the critical skills and capabilities needed to achieve its mission. This program is tied directly to the Agency's major education goal of strengthening the future STEM workforce for NASA and the Nation.

NASA's New Museum Grant Allies Will Make the Universe Accessible

In January 2010, NASA announced awards for science museums and planetariums. Interactive museum exhibits about climate change, Earth science, and missions beyond Earth are among the projects NASA selected to receive Agency funding. Nine informal education providers from Alaska to New York will share $6.2 million in grants through NASA's Competitive Program for Science Museums and Planetariums.

Participating organizations include museums, science centers, Challenger Centers, and other institutions of informal education. Selected projects will partner with NASA's Museum Alliance, an Internet-based, nationwide network of more than 400 science centers, planetariums, museums, aquariums, zoos, observatory visitor centers, NASA visitor centers, nature centers, and park visitor centers.

Projects in the program will engage learners of all ages as well as educators who work in formal or informal science education. The projects will provide NASA-inspired space, science, technology, engineering, or mathematics educational opportunities, including planetarium shows and exhibits.

In conjunction with NASA's Museum Alliance, the grants focus on NASA-themed space exploration, aeronautics, space science, Earth science, microgravity, or a combination of themes. Some projects will include partnerships with elementary and secondary schools, colleges, and universities.

The projects are located in Alaska, Colorado, Florida, Illinois, New York, North Carolina, Oregon, and South Dakota. The nine grants have a maximum 5-year period of performance and range in value from approximately $120,000 to $1.15 million. Selected projects work with the NASA Shared Service Center in Mississippi to complete the business review necessary before a NASA award is issued.

Proposals were selected through a merit-based, external peer-review process. NASA's Office of Education and Mission Directorates collaborated to solicit and review the grant applications. This integrated approach distinguishes NASA's investment in informal education. NASA received 67 proposals from 32 states and the District of Columbia.

Congress initially funded the Competitive Program for Science Museums and Planetariums grants in 2008. The first group of projects began in fall 2009 in California, Colorado, Florida, Illinois, Iowa, Michigan, Minnesota, Montana, New York, North Carolina, Vermont, and Washington. Congress enacted funds to continue this program in 2010.

Education Secretary Hosts Students for Talk with Space Station

In November 2009, NASA Administrator Charles Bolden and Secretary of Education Arne Duncan hosted Washington D.C.-area middle and high school students for a live discussion with astronauts aboard the ISS. This event was part of an annual collaboration between NASA and the Department of Education to celebrate International Education Week.

The media was invited to attend the chat between the space station's Expedition 21 crew and students from the Washington Mathematics Science Technology Public Charter High School and the Parkland Magnet Middle School for Aerospace Technology.

The call took place between 10:10 and 10:30 a.m. EST, during an event on November 5, in the auditorium of the Department of Education. As part of the 10th annual celebration of International Education Week, the students asked the crewmembers questions in English, French, German, and Russian. The week highlighted international education and international exchange,

Students at the Washington Mathematics Science Technology Public Charter High School and the Parkland Magnet Middle School for Aerospace Technology talked to the International Space Station crew of Expedition 21.

and the 2009 theme was "Creating a Vision for a Better Future."

The international Expedition 21 crew participating in the event consisted of NASA astronauts Jeff Williams and Nicole Stott, European Space Agency astronaut Frank De Winne, Canadian Space Agency astronaut Robert Thirsk, and Russian cosmonauts Roman Romanenko and Maxim Suraev. Patrick Forrester, Jose Hernandez, and Christer Fuglesang, all who recently flew on NASA's STS-128 space shuttle mission, and former astronaut Don Thomas, a veteran of four space flights, also participated.

NASA Administrator Bolden compared the goals of International Education Week with the Agency's work with 16 nations to complete and operate the ISS, which he described as "the most advanced scientific research platform ever created."

"The theme for International Education Week 2009, 'Creating a Vision for a Better Future,' echoes NASA's commitment to work toward the common goals of making life better for people here on Earth and improving our understanding of the universe," Bolden said.

Secretary of Education Duncan explained the significance of the event: "International education is an important part of the well-rounded and complete education we need to offer all of our Nation's children. Our graduates should be global citizens prepared to work well with people from diverse backgrounds, whether [an] individual who is a recent immigrant to the United States living in the community, or a business client or colleague located halfway around the world."

The downlink is one in a series with educational organizations in the United States and abroad to improve teaching and learning in science, technology, engineering, and mathematics.

> *NASA awarded grants to five minority serving institutions to support higher education teaching and learning in STEM disciplines.*

NASA Awards Education Research Grants to Minority Universities

NASA has awarded education grants to five minority serving institutions to develop innovative projects in support of higher education teaching and learning in STEM disciplines.

NASA's Minority University Research and Education Programs Small Programs project is designed to enhance students' academic experiences and encourage underserved and underrepresented groups to pursue STEM careers, which are critical to NASA's missions.

Grants were awarded to the following colleges, universities, and partnerships:

- Navajo Technical College in Crownpoint, New Mexico
- Florida Agricultural and Mechanical University in Tallahassee, Florida
- New Mexico State University in Las Cruces, New Mexico
- North Carolina Agricultural and Technical State University in Greensboro, North Carolina
- Sistema Universitario Ana G. Méndez Inc. in Caguas, Puerto Rico

The five projects will receive funding ranging from $90,800 to $345,850. They are eligible for renewal for 2 years, based on project performance and funding availability. Kennedy manages the project for the Agency.

NASA Announces Global Climate Change Education Awards

In October 2009, NASA announced education awards for global climate change. Through NASA's Global Climate Change Education initiative, 15 organizations

As Earth's climate changes, NASA is encouraging the study of this phenomenon by awarding funding to universities and organizations across the country to engage this subject using NASA's Earth observation data and Earth system models.

across the United States are receiving extra funding to help enhance students' learning potential through the use of NASA's Earth science resources.

The Challenger Center for Space Science Education in Alexandria, Virginia, was chosen as one of the recipients. The center plans to use the grant money to develop a set of engaging, interactive, learning activities that help middle school students learn about and explore climate change from an orbital perspective. The activities will be distributed for use across the 46 Challenger Learning Centers.

In total, $6.1 million in cooperative agreements was awarded to selected colleges and universities, nonprofit groups, museums, science centers, and a school district. The Global Climate Change Education program is managed by Langley Research Center.

Each cooperative agreement is expected to leverage NASA's unique contributions in climate and Earth system science. These grants support NASA's goal of engaging students in the critical disciplines of science, technology, engineering, and mathematics, and inspiring the next generation of explorers.

The winning proposals illustrate innovative approaches to using NASA content to support elementary, secondary, and undergraduate teaching and learning, and encourage lifelong learning. There is particular emphasis on engaging students using NASA Earth observation data and Earth system models.

The 15 proposals will fund organizations in 12 states: Alaska, Arizona, California, Colorado, Florida, Georgia, Mississippi, New York, North Dakota, Pennsylvania, Virginia, and Wisconsin. Winning proposals were selected through a merit-based, peer-reviewed competition. The awards have up to a 3-year period of performance and range in value from about $170,000 to $650,000.

The cooperative agreements are part of a program Congress began in fiscal year 2008.

NASA Launches New Education Initiatives with Disney's Buzz Lightyear

For Buzz Lightyear, the voyage with NASA continues. NASA and Disney Parks, which collaborated to carry toy space ranger Buzz Lightyear into orbit, are launching new efforts to encourage students to pursue studies in science, technology, engineering, and mathematics. The 12-inch-tall action figure spent more than 15 months aboard the ISS and returned to Earth on September 11, 2009. On October 2, a ticker-tape parade at Walt Disney World's Magic Kingdom in Orlando, Florida, officially welcomed Lightyear home.

NASA astronaut Mike Fincke, the station commander from October 2008 to April 2009, spent the day at the Magic Kingdom to tell students about an educational design challenge and a new online game:

NASA and Disney Parks, which collaborated to carry toy space ranger Buzz Lightyear into orbit, are launching new efforts to encourage students to pursue studies in science, technology, engineering, and mathematics.

- Mission Patch Design Challenge: Students ages 6–12 had the opportunity to design a patch to commemorate Lightyear's mission and his accomplishment of being the longest serving space ranger. The student with the most creative mission patch and 100-word essay won a tour of Kennedy Space Center in Florida and a trip to Walt Disney World Resort. NASA flew the winning patch into space aboard Space Shuttle Atlantis on mission STS-132 in May 2010, and presented it to the contest winner after its return to Earth.

- NASA and Disney Parks also launched a new online game as part of the Space Ranger Education Series. The series includes fun educational games for students and materials for educators to download and integrate into classroom curricula. In the newest game, "Putting It All Together," players can build the entire station using all of the real modules.

"We can't thank our partners at NASA enough for bringing Buzz Lightyear home from space to his family, friends, and fans here at Disney Parks—after all, this was his dream come true," said Duncan Wardle, vice president of Disney Parks. ❖

Partnership News

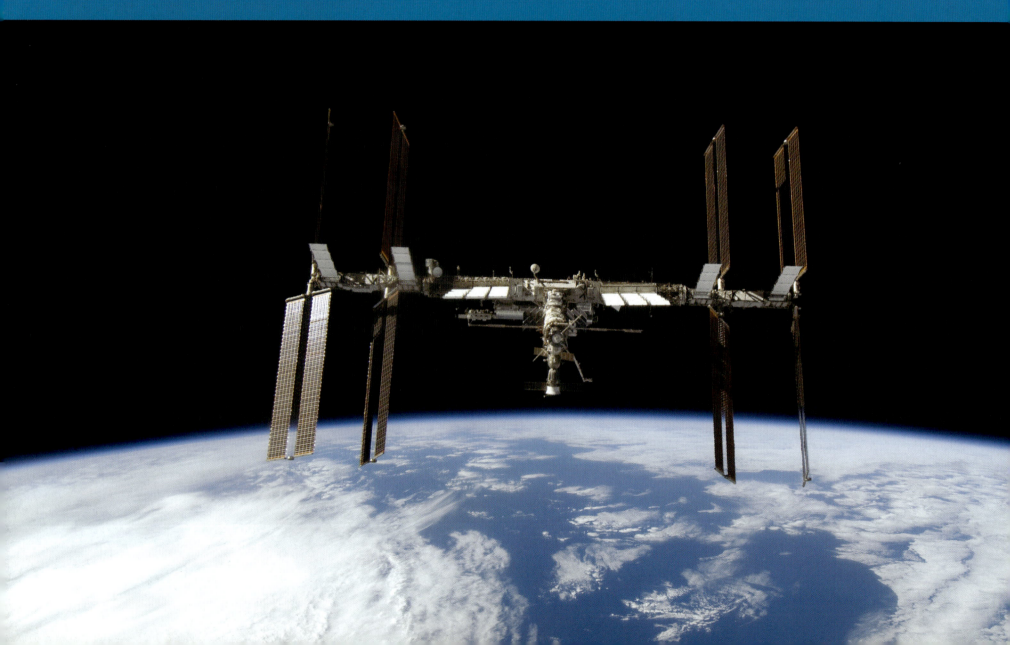

NASA enriches the lives of people everywhere through partnerships with private industry, academia, and other government agencies. Employing NASA's advanced facilities, technologies, and wealth of technical expertise, these successful partnerships are helping to combat wildfires and floods, prevent disease, encourage technological development, and provide many other benefits—bringing space-inspired science back to Earth.

Partnership News

NASA Partners to Revolutionize Personal Transportation

NASA officials have signed an agreement with Unimodal Systems Inc. to collaborate on the use of NASA-developed control software and human factors techniques to evaluate acceleration, jerk, and vibration of an advanced transportation vehicle system. The software was originally designed to control robots and other applications. The collaboration will help NASA better understand the software's usefulness, human performance, and safety.

"This collaborative effort is anticipated to help NASA with its aeronautics and space activities, while Unimodal gets to develop the next generation high-speed transportation system," said Jeffery Smith, deputy chief of the Entrepreneurial Initiatives Division at Ames Research Center. "NASA will receive valuable feedback from our systems software usage." Per the agreement, Unimodal will contribute its SkyTran vehicle, currently located at the NASA Research Park, and its advanced transportation technology; NASA will provide its Plan Execution Interchange Language (PLEXIL) and Universal Executive (UE) software to control the vehicle.

In the future, SkyTran will use small vehicles running on elevated, magnetically levitated (maglev) guideways, which distinguishes it from other railed systems. The vehicles are lightweight, personal compartments that can transport up to three passengers. Travelers board the pod-like vehicles and type their destinations into a small computer. Using intelligent control system software, SkyTran will run non-stop, point-to-point service without interrupting the flow of traffic.

These vehicles will eventually travel up to 150 mph and move 14,000 people per hour, both locally and regionally. SkyTran will serve as a feeder system to other transit systems, such as subways and high-speed rail.

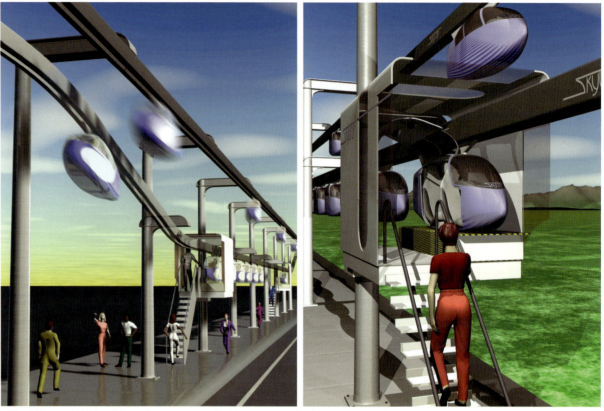

With aid from a NASA collaborative agreement, SkyTran may one day become an environmentally friendly transportation system, ferrying commuters in small, three-person pods along magnetically levitated guideways.

"SkyTran's personal rapid transit has generated serious interest with local, regional, and state transportation leaders who are considering funding the building of the Unimodal maglev PRT system in the NASA Research Park," said Michael Marlaire, director of the NASA Research Park at Ames. "This construction and new R&D partnership may usher a new 'green' technology maglev PRT system into Silicon Valley." "We're working with NASA and aerospace engineers to ensure aerospace-level standards that exceed the safety records of current transportation systems," explained Christopher Perkins, chief executive officer of Unimodal. Both organizations will mutually benefit. NASA will receive feedback on its software's usefulness in ground-based propulsion systems, while Unimodal will develop a transportation system designed to eliminate traffic congestion, mitigate greenhouse gasses, and reduce dependence on foreign oil. "For cities across the Nation, SkyTran will create green technology jobs and launch a new era of public-private partnerships that will make public transit affordable to install, and profitable to operate," said Perkins.

NASA Wind Tunnel Used to Test Truck Fuel Efficiency

Saving the Nation $10 billion annually in diesel fuel costs may be possible in a few years, thanks to new devices developed at Lawrence Livermore National Laboratory (LLNL) and now being tested at Ames.

In support of the U.S. Department of Energy's mission to reduce U.S. dependency on fossil fuels, LLNL has teamed with Ames; Navistar Inc., of Warrenville, Illinois; the U.S. Air Force; and industry to develop and test devices for reducing the aerodynamic drag of tractor-trailers. The devices could increase the trucks' fuel efficiency by as much as 20 percent.

Tractor-trailers make up about 12 percent of U.S. petroleum consumption, or 21 million barrels per day. The average fuel mileage of a tractor-trailer is 6 miles per gallon. A 2-percent reduction in the aerodynamic drag of tractor-trailers translates into 285 million gallons of diesel fuel saved per year.

"We are delighted to host this important test that could help our Nation save billions of dollars in fuel costs each year," said S. Pete Worden, director of Ames. "This is an excellent example of what can be accomplished through our collaboration with other Federal laboratories and industry."

Aerodynamic drag is caused from pressure differences around the vehicle. At highway speeds, a tractor-trailer uses more than 50 percent of the energy produced by the engine to overcome aerodynamic drag, while rolling resistance consumes roughly 30 percent of the usable energy.

Complementing 30 years of tractor-trailer aerodynamic research and development, LLNL computer simulations have identified critical drag producing regions around the trucks, such as the trailer base, the underbody, and the gap between tractor and trailer. LLNL scientists estimate that with aerodynamic devices placed in these regions, the trucking industry could see as much as a 20-percent increase in the fuel mileage efficiency rate, which saves 4.9 billion gallons of diesel per year, equaling approximately $14.7 billion in savings.

A truck trailer is prepped to be placed inside the world's largest wind tunnel, Ames Research Center's National Full-Scale Aerodynamics Complex. Testing in the facility of aerodynamic devices designed to reduce drag on tractor-trailers could ultimately result in millions of gallons of diesel fuel saved per year.

"This is a technology that could easily be installed on the tractor-trailer trucks that are out on the highway today," said Kambiz Salari, LLNL's lead scientist on the project. "And it's time to market is incredibly quick. In just 3 years, we could see these devices on the road and realize the real fuel savings."

The lab is conducting a full-scale test in the world's largest wind tunnel, the National Full-Scale Aerodynamics Complex (NFAC), which operates under the direction of the Arnold Engineering Development Center at Ames. The goal is to identify drag reduction devices, both commercially available and under development, that show the potential for improving fuel efficiency. The wind tunnel test section's huge size—80 feet by 120 feet—makes it ideal for testing a full-scale semi with a 53-foot trailer.

"This testing highlights a special opportunity for an Air Force-run facility to participate in research in areas beyond the Department of Defense and work to improve everyday issues such as fuel economy on national roadways," said Christopher Hartley, test engineer for Jacobs Engineering Group Inc., based at the NFAC.

The commercially available devices to be tested are manufactured by Aerofficient, Aeroindustries, AT Dynamics, Freight-wing, Ladyon, and Windyne.

Prototype devices currently under development will be provided by LLNL and Navistar, which are collaborating to get proven drag reduction devices on the road. Performance will be evaluated under different tractor-trailer combinations.

Livermore's project is funded by the Department of Energy's Energy Efficiency and Renewable Energy Program's Freedom Cooperative Automotive Research and fuel partnership.

> *"Making our trucks more fuel efficient means . . . we can get our goods to the general public in a more timely, and ultimately, less expensive way."*

"Making our trucks more fuel efficient means we can not only travel further using less fuel, but it means we can get our goods to the general public in a more timely, and ultimately, less expensive way," said Ron Schoon, chief engineer of aerodynamics at Navistar.

The lab is collaborating with Navistar to push the state of the art in semi-truck aerodynamics and design the next generation of highly aerodynamic, integrated, energy efficient semi-trucks. Navistar International Corp., formerly International Harvester Company, produces commercial trucks, mid-range diesel engines, and other vehicle products.

NASA Chooses Small Business High-Tech Projects for Development

NASA has selected for development 368 small business innovation projects that include research to minimize aging of aircraft, new techniques for suppressing fires on spacecraft, and advanced transmitters for deep space communications.

Chosen from more than 1,600 proposals, the competitively selected awards will address Agency research and technology needs. The awards are part of NASA's Small Business Innovation Research (SBIR) and Small Business Technology Transfer (STTR) programs.

The SBIR program selected 335 proposals for negotiation of Phase I contracts, and the STTR program chose 33 proposals for negotiation of Phase I contract awards. The selected SBIR projects have a combined value of approximately $33.5 million. The selected STTR projects have a combined value of approximately $3.3 million.

The SBIR contracts will be awarded to 245 small, high-technology firms in 36 states. The STTR contracts will be awarded to 31 small, high-technology firms in 19 states. As part of the STTR program, selected firms will partner with 26 universities and research institutions in 20 states.

Past innovations from the program have benefited a number of NASA efforts, including air traffic control systems, Earth observing spacecraft, the International Space Station, and the development of spacecraft for exploring the solar system.

A few of the research areas among this group of selected proposals include:

- Advanced aerospace adhesives to minimize aging and increase durability of aircraft
- Novel computational tools to better design future hypersonic spacecraft
- New approaches to fire suppression in spacecraft environments
- Technologies to monitor crew health and well-being using very small scale testing devices
- New instruments for small lunar rovers or landers to enable critical mineralogical analysis for studying regolith, rock, ice, and dust samples
- Advanced transmitters for deep space communications

As a highly competitive, three-phase award system, the SBIR program provides qualified small businesses—including women-owned and disadvantaged firms—with opportunities to propose unique ideas that meet specific research and development needs of the Federal government. The criteria used to choose these winning proposals included technical merit and feasibility, experience, qualifications and facilities, effectiveness of the work plan and commercial potential and feasibility.

NASA works with U.S. industry to infuse pioneering technologies into Agency missions and transition them into commercially available products and services.

NASA Develops Algae Bioreactor as a Sustainable Energy Source

As a clean energy alternative, NASA invented an algae photo-bioreactor that grows algae in municipal wastewater to produce biofuel and a variety of other products. The NASA bioreactor is an Offshore Membrane Enclosure for Growing Algae (OMEGA) and will not compete with the agriculture industry for land, fertilizer, or freshwater.

Ames licensed the patent-pending algae photo-bioreactor to Algae Systems LLC, of Carson City, Nevada, which plans to develop and pilot the technology in Tampa Bay, Florida. The company plans to refine and integrate the NASA technology into biorefineries to produce renewable energy products, including diesel and jet fuel.

"NASA has a long history of developing very successful energy conversion devices and novel life support systems," said Lisa Lockyer, deputy director of the New Ventures and Communications Directorate at Ames. "NASA is excited to support the commercialization of an algae bioreactor with potential for providing renewable energy here on Earth."

The OMEGA system consists of large plastic bags with inserts of forward-osmosis membranes that grow freshwater algae in processed wastewater by photosynthesis. Using energy from the Sun, the algae absorb carbon dioxide from the atmosphere and nutrients from the wastewater to produce biomass and oxygen. As the algae grow, the nutrients are contained in the enclosures, while

A newly licensed NASA technology will soon be using algae to cleanse water and provide a sustainable, inexpensive fuel source.

the cleansed freshwater is released into the surrounding ocean through the forward-osmosis membranes.

"The OMEGA technology has transformational powers. It can convert sewage and carbon dioxide into abundant and inexpensive fuels," said Matthew Atwood, president and founder of Algae Systems. "The technology is simple and scalable enough to create an inexpensive, local energy supply that also creates jobs to sustain it."

When deployed in contaminated and "dead zone" coastal areas, this system may help remediate these zones by removing and utilizing the nutrients that cause them. The forward-osmosis membranes use relatively small amounts of external energy compared to the conventional methods of harvesting algae, which have an energy intensive dewatering process.

Potential benefits include oil production from the harvested algae, and conversion of municipal wastewater into clean water before it is released into the ocean. After the oil is extracted from the algae, the algal remains can be used to make fertilizer, animal feed, cosmetics, or other valuable products.

This successful spinoff of NASA-derived technology will help support the commercial development of a new wastewater treatment method as well as an algae-based biofuels industry.

NASA Portable Hyperbaric Chamber Technology Finds Home on Earth

NASA has signed a patent license agreement with a California company to improve the medical community's access to hyperbaric chambers used to treat many medical conditions and emergencies. OxyHeal Medical Systems Inc., of National City, California, will develop new products based on technologies NASA originally developed for space.

Hyperbaric chambers create an environment in which the atmospheric pressure of oxygen is increased above normal levels. The high concentrations of oxygen can reduce the size of gas bubbles in the blood and improve blood flow to oxygen-starved tissues.

"These technologies will allow OxyHeal to develop new products capable of providing life-saving treatments and care to patients in remote areas that may not have access to large, fixed-site hyperbaric chamber facilities," said Ted Gurnee, president of OxyHeal. Additionally, the company is working on solutions that involve large portable hyperbaric chambers for possible use in treatment of disaster victims.

> "NASA has a long history of making space-aged technologies available for commercialization, creating new markets that power the economy."

The partially exclusive patent license agreement allows the company to use three technologies developed at NASA's Johnson Space Center in Houston that are associated with inflatable spacecraft modules and portable hyperbaric chambers.

NASA developed the technologies as part of a program to plan for how astronauts in space might be treated for decompression sickness. Decompression sickness, commonly called "the bends," can occur in astronauts as they undergo pressure changes returning from spacewalks and in divers as they return to the water's surface.

In addition to treating decompression sickness, hyperbaric chamber therapy on Earth also commonly provides treatment for carbon monoxide poisoning, crush injuries, healing-problem wounds, soft tissue infections, significant blood loss, and other ailments.

The NASA inventors of the portable hyperbaric chamber, James Locke, William Schneider, and Horacio de la Fuente, recently were recognized by the Federal Laboratory Consortium with a "Notable Technology Development Award."

"NASA has a long history of making space-aged technologies available for commercialization, creating new markets that power the economy," said Johnson's Michele Brekke. "These commercial products and services allow the taxpayers to benefit from space exploration."

Robonaut 2 will soon take up residence on the International Space Station, becoming the first humanoid robot to travel and work in space.

NASA and General Motors Create Cutting-Edge Robotic Technology

NASA and General Motors (GM) are working together to accelerate development of the next generation of robots and related technologies for use in the automotive and aerospace industries.

Engineers and scientists from NASA and GM worked together through a Space Act Agreement at Johnson to build a new humanoid robot capable of working side by side with people. Using control, sensor, and vision technologies, future robots could assist astronauts during hazardous space missions and help GM build safer cars and plants.

The two organizations, with the help of engineers from Oceaneering Space Systems, of Houston, developed and built the next iteration of Robonaut, a humanoid robot designed for space travel. The original was built 10 years ago by the Software, Robotics, and Simulation Division at Johnson in a collaborative effort with the Defense Advanced Research Projects Agency. Robonaut 2, or R2, is a faster, more dexterous and more technologically advanced robot. This new-generation robot can use its hands to do work beyond the scope of prior humanoid machines. R2 can work safely alongside people, a necessity both on Earth and in space.

"This cutting-edge robotics technology holds great promise, not only for NASA, but also for the Nation," said Doug Cooke, associate administrator for the Exploration Systems Mission Directorate at NASA Headquarters. "I'm very excited about the new opportunities for human and robotic exploration that these versatile robots will provide across a wide range of applications."

"For GM, this is about safer cars and safer plants," said Alan Taub, GM's vice president for global research and development. "When it comes to future vehicles, the advancements in controls, sensors, and vision technology can be used to develop advanced vehicle safety systems. The partnership's vision is to explore advanced robots working together in harmony with people, building better, higher quality vehicles in a safer, more competitive manufacturing environment."

The idea of using dexterous, human-like robots capable of using their hands to do intricate work is not new to the aerospace industry. During the past decade, NASA has gained significant expertise in building robotic technologies for space applications, including the original Robonaut. These capabilities will help NASA launch a bold new era of space exploration.

"Our challenge today is to build machines that can help humans work and explore in space," said Mike Coats, Johnson's director. "Working side by side with humans, or going where the risks are too great for people, machines like Robonaut will expand our capability for construction and discovery."

Flying Telescope Passes Key Test

Most astronomers would not dream of opening their observatory's doors in 100-mph winds, yet NASA's new SOFIA telescope recently flew in an airplane at 250 mph with the door wide open.

In December 2009, the Stratospheric Observatory for Infrared Astronomy (SOFIA) flew in a modified Boeing 747 at 15,000 feet for 1 hour and 19 minutes. For 2 minutes of that time, the door by the telescope was wide open.

"This was the first time the door was fully opened in flight," said Bob Meyer, SOFIA program manager at Dryden Flight Research Center. "We wanted to find out whether opening the door affected flying and handling the aircraft, caused acoustic resonance in the cavity, made anything come loose in the cavity because of wind.

"When you blow air over a soda pop bottle and hear a hum, that's acoustic resonance. If that happened in the plane, it could vibrate the structure of the plane and the telescope and cause problems."

SOFIA passed with flying colors. "Everything went well. No adjustments or corrections were needed. Nothing shook loose or got damaged."

The 98-inch infrared telescope is ultimately destined to fly at 40,000 feet and study a range of astronomical objects over its expected 20-year lifetime. Those objects include other galaxies and the center of our own Milky Way Galaxy; the interstellar medium, especially the building blocks of life it contains; the formation of stars and planets; and comets and asteroids in our solar system.

The veil of water vapor enveloping Earth acts like an invisible brick wall to the infrared energy from cosmic objects SOFIA wants to see. SOFIA solves this problem by viewing the heavens from "above the veil"—something ground-based scopes cannot do. Like space-based telescopes, SOFIA will collect infrared energy before it reaches Earth.

The SOFIA telescope peers out of the fuselage of its Boeing 747 mothership. The telescope will provide infrared imagery of a range of astronomical objects during flights at 40,000 feet—above the atmosphere's infrared-disrupting veil of water vapor.

As in the test, the telescope, with its primary, secondary, and tertiary mirrors, will sit in a cavity in the rear of the plane and look through the open door of the aircraft. The telescope's controls, computers, spectrometers, and other instruments will ride in the pressurized cabin.

More testing is planned before SOFIA can begin science operations.

"We'll test at all the speeds the plane can fly and all the altitudes planned for the mission," said Meyer. "We'll test different pointing elevations of the telescope itself.

"Our first light test, where we actually look at an image and characterize the telescope, is set for April 2011. In that test we'll unlock and uncage the telescope so it will move as though it's really observing. The wind will be buffeting and shaking SOFIA, so that will be the first true test of its ability to obtain stable images."

How do you keep a telescope still enough to point accurately and stay "on point" in a moving airplane with the door open?

"The telescope," explained Meyer, "rests on big shock mounts that isolate the mechanical vibrations of the plane from it. And on the back edge of the cavity there's a ramp that catches the airflow entering the cavity and redirects it back over the ramp and out of the cavity."

SOFIA will also have weights attached to it that can be sized and tuned to dampen any shaking. And the drive system can move the scope back and forth to compensate for lower frequency vibrations or movement of the aircraft. The secondary mirror can even be oscillated to remove the shaking of the image.

"SOFIA is really a marvelous piece of engineering," concluded Meyer. "This flight test represents a huge success and significant milestone for all the people who have worked hard for a decade on this mission."

Langley Wire Inspection Technology Wins 'NASA Government Invention of the Year'

Langley Research Center's K. Elliott Cramer, Daniel F. Perey, and William T. Yost are the co-inventors of Ultrasonic Wire Crimp Inspection Technology, which was selected as the 2009 "NASA Government Invention of the Year" winner.

The invention enables wire crimp connections to be checked for quality using a hand-held tool that measures

The 2009 "NASA Government Invention of the Year"—Ultrasonic Wire Crimp Inspection Technology—provides reliable testing of critical electrical connections.

Tim Griffin works with the mobile leak detector before a sampling flight. The shuttle leak detection system used at the launch pads is the size of three refrigerators, but Griffin's team reduced it in size and added automation so it could be mobile.

Griffin and his team of researchers carried the leak detector by hand into the cone of the Turrialba volcano in Costa Rica in 2005.

response to a signal sent through the wire joint. Current testing methods, which include a physical pull test (which can be destructive) and measuring the electrical resistance of the joint (which can be time consuming) have proven to be unreliable. The technology can be applied to many different crimping operations, especially critical connections such as those employed on flight vehicles.

The technology is being used as part of the Aircraft Aging and Durability Project to investigate failure of electrical wiring systems in commercial and military aircraft.

Shuttle Leak Detector at Heart of Volcano Alert System

As Tim Griffin and his team were working on better ways to detect hazardous gasses on the shuttle launch pad, they found out they also could build something to find hazardous gasses venting from a volcano. That means they may be close to building an early warning system for volcano eruptions—a system that could give those near an active cone days or more to evacuate to safety.

"There are all kinds of volcano eruptions, some have all kinds of gasses and some don't have any gasses," Griffin said. "The long-term idea is that we'd be able to characterize the volcanoes. Then if the volcano becomes more active, we can get a better idea of what's going on, how active it is, [and] do we think it's going to be a violent eruption or mainly gasses coming out?"

Griffin, who is the chief of Kennedy Space Center's Chemical Analysis Branch and holds a Ph.D. in chemistry, never studied volcanoes. Instead, his group's goal was to shrink the leak detection system at the launch pad from the size of three refrigerators to something that could be carried by hand, in a car, or perhaps inside a spacecraft.

"This project started off as a way to push the boundaries with our shuttle system," said Richard Arkin of ASRC Aerospace, the detector's co-designer. "We wanted to make it smaller, more powerful and lighter while still maintaining operational abilities and maintenance."

Parts of the miniaturization work were easy, such as going from numerous sampling ports required at the pad to a single port for the smaller machine. Other aspects, such as building smaller pumps and additional components, required innovation and invention. In both, a mass spectrometer is used to find out what chemicals are present in the air.

They also set out to make the unit relatively autonomous, but still reliable and hearty.

At this point, the detector weighs in at 75 pounds. It stands about 9 inches tall, and its footprint is a bit larger than a backpack. In fact, one of the goals of the project is to make it small enough to be carried in a backpack.

Griffin was talking about some of the work involved in chemical analysis at a conference when officials from Costa Rica's scientific program asked about applying the technology to the volcanic studies. It started to look like a natural fit.

Costa Rica proved a good testing ground for the equipment because most of the population lives around or near four active volcanoes. They not only worry about sudden eruptions, but also high concentrations of carbon dioxide the volcanoes vent. The gas tends to kill all vegetation and livestock near the venting areas, but people cannot see the carbon dioxide.

The detector showed a way to find out where the gas pockets are and how they change. The team flew the detector on three different kinds of airplanes, where it modeled the chemicals in volcanic plumes in three dimensions, a level of precision that astonished Arkin.

"That was something that I never thought about doing," Arkin said.

The team put the detector in the backseat of a car and drove it through Costa Rican cities to sample the air and also carried it into the volcanoes by hand. In the future, Griffin wants to load it inside drones so the detection system can fly directly into the plumes of erupting mountains without endangering a pilot.

The results are expected to provide more information to help researchers pinpoint what volcanoes are doing at any given time, and when or if they might be about to spew.

Although the highest potential is still a few years away for the detection system, Griffin said he can envision a time when there are a number of detectors based around the world ready to scan volcanoes suspected of erupting. The extra information could be enough to convince officials to order an evacuation before it's too late.

Two NASA Technologies Inducted into the Space Technology Hall of Fame

Lenses that protect human eyes from solar radiation and digital fly-by-wire technology that significantly improves the fuel efficiency, reliability, and cost of modern aircraft have been selected for induction into the

Inducted this year into the Space Technology Hall of Fame, Eagle Eyes Optics sunglasses filter out harmful wavelengths of light known to contribute to cataracts and other eye conditions.

Space Foundation's Space Technology Hall of Fame. This recognition honors outstanding technologies developed for use in space that have been adapted to improve life on Earth.

Eagle Eyes Optics and NASA's Jet Propulsion Laboratory (JPL) will be inducted for radiation-filtering technology now used in sunglasses, and Venture Technology Inc. will receive a commendation for its early work on the optical technology.

Dryden, Draper Laboratory, The Boeing Company, and Airbus will be inducted for developing a technology that uses electronic pulses instead of mechanical or hydraulic linkages to control the actuators for the ailerons, flaps, and other control surfaces of an aircraft.

In the 1960s, JPL began looking for a method to protect human eyesight from the harmful effects of solar radiation in space, especially ultraviolet and blue-light rays that are known to contribute to cataract and age-related macular degeneration.

Researchers learned that the eyes of birds-of-prey contain oil droplets that filter out harmful radiation and permit only specific wavelengths of light to enter. This natural filtering process provides them with extreme visual clarity—even at great distances. The researchers replicated this natural system and created the first transparent welding curtain to filter out harmful light and protect human eyesight in hazardous environments.

The new technology was soon applied to sunglasses and introduced to the public under the SunTiger name, known today as Eagle Eyes Optics. These lenses absorb over 99 percent of all photo wavelengths considered hazardous to human eye tissue—including ultraviolet and blue light up to 475 nanometers in the wavelength spectrum, the cut-off point at which visible light is allowed to transmit through the lens. This range is the most beneficial to the human eye for protection, for increased visual perception and contrast, and for reduced chromatic aberration.

"Eagle Eyes products are recognized around the world for their protective and vision-enhancing capabilities," said Kevin Cook, the Space Foundation's director of space awareness programs, who administers the Space Technology Hall of Fame. "They are another excellent example of how space technology benefits life on Earth."

In the late 1960s, engineers at NASA Flight Research Center (now Dryden Flight Research Center) proposed replacing bulky mechanical flight-control systems on aircraft with much lighter-weight and more reliable fly-by-wire technology. The engineers had been exploring ways to apply knowledge gained from their work in support of the Apollo Program.

They initially envisioned controlling aircraft using analog computers—the standard of the era; however, astronaut Neil Armstrong, who had successfully flown to the Moon using digital technology developed for the Apollo Program, encouraged Dryden to explore a digital option. Dryden teamed up with Draper Laboratory, which had worked with NASA on the Apollo Program, to develop a digital fly-by-wire (DFBW) solution using the specialized hardware and software developed for Apollo.

This modified F-8 Digital Fly-By-Wire Crusader was the test bed of the fly-by-wire flight control systems now used on the space shuttles and military and civil aircraft to make them safer, more maneuverable, and more efficient. Digital fly-by-wire and its developers were inducted into the Space Technology Hall of Fame this year.

Proven performance and solid cooperation between NASA and industry translated into use of DFBW systems in new aircraft design in a remarkably short time. Airbus and Boeing developed the first commercial airliners using the new DFBW technology.

DFBW-equipped aircraft are more fuel efficient because they maintain constant speed and altitude over long distances. Without bulky hydraulics, cables, rods, and pulleys, they have greater payload capacity and greater range. The electronic systems require less maintenance and, because there are fewer mechanical parts to fail, the airplanes are far more reliable. DFBW systems are easier to install and troubleshoot, which makes assembly and maintenance more efficient, and the reduced vulnerability to battle damage makes DFBW ideal for military aircraft.

"DFBW technology is a quantum jump in aircraft design and performance," said Cook. "Safer and more efficient aircraft around the globe represent a major aeronautics success, as well as another example of how technology developed for space exploration directly benefits life on Earth."

The Space Technology Hall of Fame was created in 1988 by the Space Foundation, in cooperation with NASA, to increase public awareness of the benefits resulting from space exploration programs and to encourage further innovation. To date, the Space Foundation has inducted 61 technologies as well as honored the organizations and individuals who transformed space technology into commercial products that improve the quality of life for all humanity.

NASA Ames 'Tops Out' First Building in 30 Years

"Beam me up!" was the message signed on the final beam hoisted into place on the iron skeleton of NASA's new building, called Sustainability Base, on Friday, March 12, 2010.

Although not yet completed, Sustainability Base has begun ushering in a new era of innovation, good will, and renewed American tradition. Under construction at Ames, the building advances the standard for what it means to be "green." Sustainability Base is expected to achieve a platinum rating under the Leadership in Energy and Environmental Design (LEED) standards for the environmentally sustainable design, construction, and operation of buildings. The building, however, goes beyond LEED to serve as a showcase of NASA and partner ingenuity, incorporating technologies designed for space exploration and applied to improve life here on our home planet. Sustainability Base will be a window to the future on Earth.

"It will be one of the greenest and highest performance buildings in the Federal government," said Steve Zornetzer, associate administrator of Ames. "Today is a good day to celebrate. It's a good day to stop, reflect, and show appreciation for work that was well done."

Together, NASA and Swinerton Builders workers and management signed the final beam as part of a celebration, called the "Topping Out." No one really knows how or when it originated, but the tradition places an evergreen tree, a flag, or both on the last beam as it is lifted into place to signify the structure has reached its height and that the skeleton is completed.

"As a company, we are proud to be part of a green effort that is so successful," said Dan Beyer, vice president of Swinerton Builders. "The tree signifies new growth as the building construction comes to fruition and is used over time; the flag represents who we are as Americans."

With NASA innovations and intelligence integrated throughout, the Ames Sustainability Base will be among the greenest and best performing buildings in the Federal government.

Over the years, the Topping Out custom remains important to ironworkers in the steel construction industry. For some, the evergreen symbolizes the successful completion of construction without loss of life; for others, it's a good luck charm for the occupants. Similarly, the flag also has multiple meanings: the construction of a Federal building, patriotism, or the American dream. Whatever the interpretation, it welcomes the future while providing a link with the past.

Honey Bees Turned Data Collectors Help Scientists Understand Climate Change

Estimates are that there are somewhere between 6 and 10 million species of insects on the planet, yet few are as charismatic as the honey bee.

Part of an order of winged insects called Hymenoptera, honey bees are best known for being prodigious producers of honey, the sweet amber substance they produce by partially digesting and repeatedly regurgitating the sugar-rich nectar found within the petals of flowering plants. They're also the workhorses of the modern industrial agricultural system, relied upon to pollinate crops ranging from almonds to watermelons to peaches. And they're even noted dancers capable of performing an array of complex "waggle" dances to communicate.

And now—thanks to an innovative project conceived by Wayne Esaias, a veteran oceanographer at NASA's Goddard Space Flight Center—bees have yet another role: that of climate data collectors.

When honey bees search for honey, colony scouts tend to scour far and wide and sample the area around a hive remarkably evenly, regardless of the size of the hive. And that, Esaias explained, means they excel in keeping tabs on the dynamics of flowering ecosystems in ways that even a small army of graduate students cannot.

The key piece of data bees collect relates to the nectar flow, which in the mid-Atlantic region tends to come in a burst in the spring. Major nectar flows, typically caused by blooms of tulip-poplar and black locust trees, leave an unmistakable fingerprint on beehives—a rapid increase in hive weight sometimes exceeding 20 pounds per day. When a nectar flow finishes, the opposite is true: hives start to lose weight, sometimes by as much as a pound a day.

By creating a burgeoning network of citizen scientists (HoneyBeeNet) who use industrial-sized scales to weigh their hives each day, Esaias aims to quantify the dynamics of nectar flow over time. Participating beekeepers send their data to Esaias, who analyzes it and posts nectar flow trend graphs and other environmental data for each collection site on HoneyBeeNet's Web page.

The size of HoneyBeeNet, which relies almost entirely on small-scale backyard beekeepers, has doubled over the last year and now includes more than 87 data collection sites. While the majority of sites are in Maryland, HoneyBeeNet now has sites in more than 20 states.

Beekeepers around the country are collecting crucial ecological data through an unusual method: weighing their hives each day. The effort was launched by a Goddard Space Flight Center researcher.

Data from the network, when combined with additional data that reach back to the 1920s, indicate that the timing of spring nectar flows have undergone extraordinary changes. "Each year, the nectar flow comes about a half-day earlier on average," said Esaias. "In total, since the 1970s, it has moved forward by about month in Maryland."

Esaias and Goddard colleague Robert Wolfe recently compared nectar flow data from HoneyBeeNet to satellite data that measures the annual "green up" of vegetation in the spring, one of the first times that scientists have attempted such a comparison. They corresponded nearly perfectly, confirming the usefulness of the citizen-science derived data from HoneyBeeNet to address changes in nectar flows.

Esaias' research offers hints about how bees might respond to climate change.

What's to blame for the remarkable warming trend in Maryland? Washington's growth has certainly played a role. Urban areas, explained Esaias, produce a "heat island" effect that causes temperatures in surrounding areas to creep upward. But, in addition to that, Esaias suspects that climate change is also contributing. And that has him nervous. "A month is a long time. If this keeps up, and the nectar flows continue to come earlier and earlier, there's a risk that pollinators could end up out of sync with the plant species that they've pollinated historically," Esaias said.

Esaias is not the only researcher who's looking at this issue. The National Academies of Science published a landmark report in 2007 that highlighted the precarious status of pollinators in North America.

Many pollinators—ranging from honey bees, to bumblebees, to lesser known species—seem to be in the midst of protracted population declines. Managed honey bee colonies, for example, have seen their numbers fall from about 5.9 million in 1947 to just 2.4 million in 2005.

In most cases, it is not clear what's causing the population declines or whether climate change is exacerbating the problem, though many researchers suspect that new types of viruses, mites, and other parasites and pesticides are important factors.

"But it's not just the honey bees that we need to be looking at," said May Berenbaum, an ecologist at the University of Illinois at Urbana-Champaign and the lead author on the National Academies report. "For honey bees, at least we can truck them around or feed them when there's a problem. It's the wild species of pollinators that are the greatest cause for concern."

Bumblebees, wasps, butterflies, and countless other insects—as well as some bats and birds—are the glue that keeps many wild ecosystems intact through pollination. And scientists are only beginning to comprehend the potential consequences that could unfold if the pollinators and the plants that rely on them get so far out of sync that extinctions begin to occur.

"To borrow an old analogy that Paul Ehrlich often used, with the wild pollinators, losing a species is a bit like losing screws in a plane," said Berenbaum. "If you lose a few here or there, it's not the end of the world, and your plane can still fly. But if you lose too many, at some point, the whole plane can suddenly come apart in mid-flight."

Indeed, entomologists have hardly begun the task of identifying wild pollinators, not to mention determining definitely which species are threatened or how they might respond as the climate shifts. Esaias' research offers hints about how bees might respond to climate change. Still, scientists estimate that there are more than 30,000 different bee species alone, and only about half of them have been formally described.

HoneyBeeNet is one way that citizens can help scientists better understand how climate change is affecting one species of pollinator. Alice Parks, a backyard beekeeper from West Friendship, Maryland, has participated for 2 years. She bought a used scale for just $26 at an auction, and weighs her hive every night.

"Weighing can be a chore sometimes," she said. "But it's such an incredibly rewarding project that it's worth it. I'm learning so much about my bees that's making me a better beekeeper, but I'm also contributing to a larger project that's helping scientists address environmental problems on a global scale."

Glenn Research Center Receives Three 'R&D 100' Awards

Three research teams at NASA's Glenn Research Center have been recognized by the editors of *R&D Magazine* with a prestigious "R&D 100" award.

The "R&D 100" award is given to the top 100 most technologically significant products of the year. The awards have been helping companies provide the important initial push a new product needs to compete successfully in the marketplace. It provides a mark of excellence known to industry, government, and academia as proof that the product is one of the most innovative ideas of the year.

The three Glenn award winners are OTIS4, the Mini-Classifier, and the L-3 ETI Model 2300HE.

The Optimal Trajectories by Implicit Simulation version 4 software, or OTIS4, is a general purpose program used to perform trajectory performance studies. Its principal application includes the preliminary design of aerospace vehicles. The trajectory performance is linked to the physical design of the vehicle by factors such as weight, fuel tank volume, and solar array sizing. OTIS4 also received NASA's 2008 "Software of the Year" award and received special recognition in May 2009 at the Northeast Ohio Software Association annual "Best of Tech" award in the category of Best Software Product. Members of the team are John P. Riehl, Waldy K. Sjauw, and Robert D. Falck of Glenn's Mission Design and

Analysis Branch, along with Stephen Paris from Boeing Research and Technology.

The Mini-Classifier is a compact low-power device that measures size distributions of aerosols, including ultrafine components less than 100 nanometers (about 4 millionths of an inch), which are of great interest to epidemiology and respiratory health. This development represents a significant advancement in particle measurement technology. Members of the team are Glenn's Paul Greenberg of the Combustion and Reacting Systems Branch; Patrick Spanos and Al Blaze of the Machining Branch; William Yanis with the National Center for Space Exploration Research in Cleveland; and Da-Ren Chen and Chaolong Qi with Washington University in St. Louis.

The L-3 ETI Model 2300HE is a high-efficiency space traveling wave tube amplifier for NASA's Lunar Reconnaissance Orbiter spacecraft. This high-efficiency, high-reliability microwave power amplifier is capable of transmitting science data and images from the Moon to Earth faster than previously possible. It is the first high data rate K-band transmitter to fly on a NASA spacecraft. Members of the team are Rainee Simons of Glenn in collaboration with Paul Spitsen, William Menninger, Neal Robbins, Daniel Bibb, and Phillip Todd of L-3 Communications Electron Technologies Inc.

Stennis All-Hazards Network Being Adopted Throughout NASA

It has been called "the wave of the future," and now the "all-hazards network" system (HazNet) developed at Stennis Space Center, is being implemented across the Space Agency.

HazNet incorporates maps, reports, Internet-derived data, and real-time sensor input into a geographic information system (GIS)-based display to provide organizations and officials with comprehensive information during emergency and disaster situations. It also allows organizations and officials to communicate, collaborate, and share data during such events, enabling a coordinated response.

"The system is a real benefit for managing incidents," Stennis' emergency director Ron Magee said. "It draws information from a variety of sources and allows you to have at your fingertips information that you need to properly respond to events."

Based on 5 years of research and thousands of hours of testing, the HazNet system was developed by NVision Solutions Inc., a Bay St. Louis, Mississippi, company that worked with NASA, the U.S. Department of Homeland Security, the Federal Emergency Management Agency, and others on the project. Most of the funding for development came through NASA programs, including the SBIR program.

"This tool ushers in a new era of information gathering and sharing between Federal, state, and local agencies across multiple geographies," said Craig Harvey, chief operating officer for NVision. "We view NASA as a key partner in its development."

Through the NASA partnership, the HazNet system was implemented in neighboring St. Tammany Parish, Louisiana, and Hancock County, where Stennis is located. A contract was awarded last year to install the system at three NASA locations: Stennis, Michoud Assembly Facility in New Orleans, and NASA Headquarters in Washington, D.C. At Stennis, the system was unveiled with the opening of the facility's new Emergency Operations Center.

An SBIR Phase III contract has been awarded to install the system throughout NASA centers, and Stennis will host emergency operations personnel from those facilities for a training session. "The system provides facility information, weather information, and various data from a number of sources," Magee explained. "It provides emergency responders with all the information they might need, all in one place."

The goal is to provide a real-time common operating picture for responders, so breakdowns in communication and gaps in knowledge will not hamper emergency and/or disaster response. At Stennis, a major feature of the system is to enable effective response during a possible hurricane; however, the system is designed to help officials make timely, informed decisions in all types of emergencies.

> "The system provides emergency responders with all the information they might need, all in one place."

For instance, during a flood, HazNet helps officials determine where flood waters are headed, allowing for warnings to be issued and thus protecting lives and property. The system also enables timely response during events such as fires, hazardous spills, and terror attacks.

Another key feature of the new system is its adaptability. As Harvey explained, features are continually being added. For instance, 3-D building displays are being incorporated, as well as emergency vehicle tracking and storm evacuation capabilities and an emergency shelter management system.

An NVision goal is to incorporate HazNet in all of the parishes and counties surrounding Stennis. However, even before that is done, there is little doubt in Magee's mind about HazNet's value. "This represents a very big step forward in safety and security for Stennis and all of NASA," he said. "We're excited to have been a part of its development."

NASA Honors 2009 Centennial Challenges Winners

In February, NASA honored the achievements of the 2009 Centennial Challenges prize winners and competition hosts with a technical symposium and recognition ceremony at NASA Headquarters in Washington, D.C. Centennial Challenges is NASA's program of technology prizes for the citizen-inventor. The program's goals are to drive progress in aerospace technology that is of value

Paul's Robotics (above) won the Regolith Excavation Centennial Challenge with its innovative robot, Moonraker (above and left), capturing the $500,000 first prize. Overall, Centennial Challenge winners garnered $3.65 million in combined prize money while developing remarkable technologies that may soon find use for space exploration or on Earth.

to NASA's missions; encourage participation of independent teams, individual inventors, student groups, and private companies of all sizes in aerospace research and development; and find innovative solutions to technical challenges through competition and cooperation.

Five Centennial Challenge events were held in 2009, awarding $3.65 million in combined prize money at four of the competitions: Regolith Excavation, Lunar Lander, Power Beaming, and Astronaut Glove.

Paul's Robotics, a team led by college student Paul Ventimiglia of Worcester Polytechnic Institute in Worcester, Massachusetts, won the $500,000 first prize in the Regolith Excavation Challenge hosted at NASA Ames Research Park and managed by the California Space Education and Workforce Institute. Other teams won the $150,000 second-place and $100,000 third-place prizes. Teams were required to design, build, and operate a mobile robot that could dig up and deposit at least 150 kilograms of material from a simulated lunar surface and deposit it in a collection bin, demonstrating a key task in lunar construction and resource harvesting.

Masten Space Systems, of Mojave, California, won the $1 million first prize in the Northrop Grumman Lunar Lander Challenge. Managed by the X PRIZE Foundation, the challenge involved building and flying a rocket-powered vehicle that simulates the flight of a vehicle on the Moon, requiring exacting control and navigation, as well as precise control of engine thrust, all done automatically. Armadillo Aerospace, of Rockwall, Texas, claimed the $500,000 second prize.

The Power Beaming Challenge, hosted at Dryden and managed by The Spaceward Foundation, was a demonstration of wireless power transmission in which teams build and demonstrate systems to beam energy from the ground to a robotic device that climbs a vertical cable. NASA is interested in power-beaming technology for purposes such as remotely powering rovers and other instruments. On Earth, the technology might supply communities with power following natural disasters, and there are intriguing applications including power beaming for airships, satellites, and space transportation, including the space elevator concept. LaserMotive LLC, of Seattle, met the stringent requirements to win the challenge's first level, winning a $900,000 prize. A prize purse of $1,100,000 remains for the next Power Beaming competition.

Peter Homer, of Southwest Harbor, Maine, winner of the 2007 Astronaut Glove Challenge, also won the 2009 challenge to develop innovative space suit glove designs to reduce the effort needed to do work during spacewalks. Homer won $250,000 for his first place design, while Ted Southern, of Brooklyn, New York, won the $100,000 second prize. The competition was managed by Volanz Aerospace.

NASA has plans for additional competitions, including the Green Aircraft Challenge set for 2011. The Power Beaming Challenge will continue, as will the Strong Tether Challenge, which is awaiting its first winner.

"I think the Centennial Challenges are very cool and exciting. Ultimately, in the real world, it can lead to things like startup companies," said Regolith Excavation Challenge winner Ventimiglia. He has plans for launching his own robotics company with the support of the prize money, and Astronaut Glove Challenge winner Homer has already launched Flagsuit LLC to explore the commercial applications of his winning technology. These are just the early signs of the impact the Centennial Challenges have beyond NASA.

According to NASA Centennial Challenges program manager Andrew Petro, "NASA's investment is relatively small, and the return is enormous. In addition to specific innovations that might be considered in future designs, the NASA engineers present were able to observe and evaluate a collection of working prototypes that would normally cost millions of dollars to create. Furthermore, there is tremendous value in giving a new generation of innovators the opportunity to test their ideas in real-world conditions, or in this case, out-of-this-world conditions. The experience they gain through these projects will aid them in their future work and benefit our Nation's economy for years to come." ❖

Space Technology Hall of Fame® is a registered trademark of the Space Foundation.

Armadillo Aerospace successfully met the Level 2 requirements for the Lunar Lander Centennial Challenge and qualified to win a $1 million first place prize. To qualify for the Level 2 prize, Armadillo Aerospace's rocket vehicle took off from one concrete pad, ascended horizontally, then landed on a second pad that featured boulders and craters to simulate the lunar surface. The team ultimately came in second place in the overall competition behind Masten Space Systems.

NASA Technology Award Winners

Since its inception in 1976, *Spinoff* has featured myriad award-winning technologies that have been recognized by NASA and industry as forerunners in innovation. Here is a chronology of these winners, including the year(s) they were featured in *Spinoff* and the year they were awarded one (or more) of the following:

R&D 100
The R&D 100 Awards were established by *R&D Magazine* in 1963 to pick the 100 most technologically significant new products invented each year. For 48 years, the prestigious R&D 100 Awards have been helping provide new products with the needed recognition for success in the marketplace.

Space Technology Hall of Fame
The Space Foundation and NASA created the Space Technology Hall of Fame in 1988. The award recognizes the life-changing technologies emerging from America's space programs; honors the scientists, engineers, and innovators responsible; and communicates to the American public the significance of these technologies as a return on investment in their Space Program.

NASA Invention of the Year
Since 1990, through the NASA Invention of the Year Award, the NASA Office of the General Counsel, in partnership with the NASA Inventions and Contributions Board, have recognized and rewarded inventors of exceptional, cutting-edge NASA technologies that have been patented in the United States.

NASA Software of the Year
Established in 1994, the NASA Software of the Year Award is given to those programmers and developers who have created outstanding software for the Agency.

Spinoff Year(s)	Technology	R&D 100	Space Technology Hall of Fame	NASA Invention of the Year	NASA Software of the Year
1976, 1977, 1981, 2002, 2005	Memory Foam		1998		
1976	Improved Firefighter's Breathing System		1988		
1977, 1979, 1983, 1989, 2004	Liquid-Cooled Garments		1993		
1978, 1990, 2009	Fabric Roof Structures		1989		
1978, 1981, 1983, 2005	Phase-Insensitive Ultrasonic Transducer	1978			
1981	Cordless Tools		1989		
1983, 1986	PMR-15 Polyimide Resin	1977	1991		
1984, 1996	Scratch Resistant Lenses		1989		

Spinoff Year(s)	Technology	R&D 100	Space Technology Hall of Fame	NASA Invention of the Year	NASA Software of the Year
1985, 2008	Redox Energy Storage System	1979			
1985	Safety Grooving		1990		
1985, 1986, 1987, 1991, 1993, 2003, 2008	Earth Resources Laboratory Applications Software (ELAS)		1992		
1987, 1989, 1993, 2006, 2010	Eagle Eyes Sunglasses		2010		
1988, 2006, 2010	Radiant Barrier		1996		
1988	Sewage Treatment With Water Hyacinths		1988		
1989	Data Acquisition and Control System Model 9450/CAMAC	1986			
1990	NASA Structural Analysis (NASTRAN) Computer Software		1988		
1991	Heart Defibrillator Energy Source		1999		
1992	Dexterous Hand Master (DHM)	1989			
1994	Ballistic Electron Emission Microscope (BEEM)	1990			
1994, 2008	Omniview Motionless Camera	1993			
1994	Stereotactic Breast Biopsy Technology		1997		
1995	1100C Virtual Window	1994			
1995, 2006, 2008	Microbial Check Valve		2007		
1996	Anti-Shock Trousers		1996		

Spinoff Year(s)	Technology	R&D 100	Space Technology Hall of Fame	NASA Invention of the Year	NASA Software of the Year
1996, 2004	Tetrahedral Unstructured Software System (TetrUSS)				1996, 2004
1996	Automated Hydrogen Gas Leak Detector	1995			
1996	Ceramics Analysis and Reliability Evaluation of Structures/Life (CARES/Life)	1995			1994
1996	Memory Short Stack Semiconductor	1994			
1997, 2004, 2009	Outlast Smart Fabric Technology		2005		
1997	Foster-Miller Fiber Optic Polymer Reaction Monitor	1990			
1997, 2007	Power Factor Controller		1988		
1997	Reaction/Momentum Wheel: Apparatus for Providing Torque and for Storing Momentum Energy			1998	
1998, 2001, 2008, 2009, 2010	Flexible Aerogel Superinsulation	2003			
1998	Data Matrix Symbology		2001		
1999, 2010	Active Pixel Sensor		1999		
1999	DeltaTherm 1000	1994			
1999	Superex Tube Extrusion Process	1995			
1999, 2010	Precision GPS Software System		2004		2000
1999	Process for Preparing Transparent Aromatic Polyimide Film			1999	
1999	Tempest Server	1999			1998
2000	Genoa: A Progressive Failure Analysis Software System	2000			1999

Spinoff Year(s)	Technology	R&D 100	Space Technology Hall of Fame	NASA Invention of the Year	NASA Software of the Year
2000	Humanitarian Demining Device		2003		
2001	Composite Matrix Resins and Adhesives (PETI-5)			1998	
2001	LARC PETI-5 Polyimide Resin	1997		1998	
2001	The SeaWiFS Data Analysis System (SeaDAS)				2003
2001	TOR Polymers	2000			
2001	Video Image Stabilization and Registration (VISAR)		2001	2002	
2002	Automatic Implantable Cardiovertor Defibrillator		1991		
2002	DeBakey Rotary Blood Pump (Ventricular Assist Device-VAD)		1999	2001	
2002	Hybrid Ice Protection System	2003			
2002	Quantum Well Infrared Photodetector (QWIP)		2001		
2002	Virtual Window		2003		
2003	Cart3D				2002
2003	Cochlear Implant		2003		
2003	Generalized Fluid System Simulation Program				2001
2003	LADARVision 4000		2004		
2003	MedStar Monitoring System		2004		
2003	Personal Cabin Pressure Altitude Monitor and Warning System			2003	

Spinoff Year(s)	Technology	R&D 100	Space Technology Hall of Fame	NASA Invention of the Year	NASA Software of the Year
2003, 2007, 2010	Portable-Hyperspectral Imaging Systems		2005		
2004	InnerVue Diagnostic Scope System		2005		
2004	Land Information System (LIS) v. 4.0				2005
2004, 2009	NanoCeram Superfilters	2002	2005		
2004	Numerical Propulsion System Simulation (NPSS)				2001
2004	Photrodes for Electrophysiological Monitoring	2002			
2005	PS/PM300 High-Temperature Solid Lubricant Coatings	2003			
2005	iROBOT PackBOT Tactical Mobile Robot		2006		
2005, 2008	Light-Emitting Diodes for Medical Applications		2000		
2005	Numerical Evaluation of Stochastic Structures Under Stress (NESSUS)	2005			
2005	THUNDER Actuators	1996			
2005, 2010	Zero-Valent Metal Emulsion for Reductive Dehalogenation of DNAPL-Phase Environmental Contaminants (EZVI)		2007	2005	
2006	Novariant RTK AutoFarm AutoSteer		2006		
2006	Petroleum Remediation Product		2008		
2007	Advanced Lubricants		2000		
2007	Atomic Oxygen System for Art Restoration	2002			
2007	Future Air Traffic Management Concepts Evaluation Tool (FACET)				2006

Spinoff Year(s)	Technology	R&D 100	Space Technology Hall of Fame	NASA Invention of the Year	NASA Software of the Year
2007	Macro-Fiber Composite Actuator	2000		2007	
2007	ResQPOD: Circulation-Enhancing Device		2008		
2007	ArterioVision: Noninvasive Cardiovascular Disease Detection		2008		
2008	DMBZ-15 High-Temperature Polyimide	2003			
2008	LaRC-SI: Soluble Imide	1995			
2008	Wireless Measurement Acquisition System	2006			
2008	Optical Backscatter Reflectometer	2007			
2009	Cryo-Tracker Probe	2001			
2009	PETI-330 Resin			2008	
2009	FPF-44 Polyimide Foam			2007	
2009	RP-46 Polyimide Resin	1992		2004	
2010	Thermal and Environmental Barrier Coatings	2007			
2010	Subaperture Stitching Interferometer	2008			
2010	Hilbert-Huang Transform	2001		2003	
2010	Inflatable Satellite Communication System	2010			
2010	Low Plasticity Burnishing (LPB)	2010			

Office of the Chief Technologist

The Office of the Chief Technologist collaborates with industry, academia, and other sources, forming productive partnerships that develop and transfer technology in support of national priorities and NASA's missions. The programs and activities resulting from these partnerships engage innovators and enterprises to fulfill NASA's mission needs while generating advancements that improve daily life.

Office of the Chief Technologist

NASA's Chief Technologist serves as the NASA Administrator's principal advisor and advocate on matters concerning Agency-wide technology policy and programs. The Office of the Chief Technologist (OCT) is responsible for direct management of NASA's Space Technology programs and for coordination and tracking of all technology investments across the Agency. The office also serves as the NASA technology point of entry and contact with other government agencies, academia, and the commercial aerospace community. The office is responsible for developing and executing innovative technology partnerships, technology transfer and commercial activities, and the development of collaboration models for NASA.

OCT will provide a technology and innovation focus for NASA through the following goals and responsibilities:

- Principal NASA advisor and advocate on matters concerning Agency-wide technology policy and programs

- Up and out advocacy for NASA research and technology programs, communication and integration with other Agency technology efforts

- Direct management of Space Technology Programs

- Perform strategic technology integration and coordination of technology investments across the Agency, including the mission-focused investments made by the NASA mission directorates

- Change culture towards creativity and innovation at NASA centers, particularly in regard to workforce development

- Document, demonstrate, and communicate societal impact of NASA technology investments and lead technology transfer and commercialization opportunities across the Agency

NASA's Space Technology initiative managed by OCT will develop and demonstrate advanced space systems concepts and technologies enabling new approaches to achieving NASA's current mission set and future missions not feasible today. This approach is in contrast to the mission-focused technology development activities within the NASA Mission Directorates, which "pull" technology development based on established mission needs. OCT and the Space Technology initiative will perform "push" technology development and demonstration. Such technologies are either crosscutting, which serves multiple NASA Mission Directorates, industry, and other government agencies; or game-changing, which enables currently unrealizable approaches to space systems and missions. OCT and Space Technology will complement the technology development activities within NASA's Mission Directorates, leveraging synergies between them, and delivering forward-reaching technology solutions for future NASA science and exploration missions and significant national needs.

In continuing to build upon the successes of NASA's former Innovative Partnerships Program (IPP), the newly established OCT is responsible for managing technology partnership development, transfer, and innovation through investments and partnerships with industry, academia, government agencies, and national laboratories.

Dr. Robert D. Braun, NASA's Chief Technologist, serves as the principal advisor and advocate on matters concerning Agency-wide technology policy and programs. Braun will help develop a broadly focused advanced concepts and technology development program leading to new approaches to future NASA missions and solutions to significant national needs.

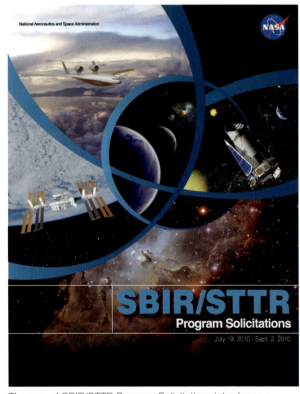

The annual *SBIR/STTR Program Solicitations* introduces a number of R&D topics and subtopics, consistent with NASA's stated needs or missions.

NASA Administrator Charles F. Bolden speaks with teachers and middle school students during the kick off of NASA's Summer of Innovation program at the Jet Propulsion Laboratory.

competitions. All of the winners participated in a technical symposium and recognition ceremony at NASA Headquarters in February. Preparations are underway for the 2011 Power Beaming and Green Flight Challenges as well as three new challenges announced this year: Nano-Satellite Launch, Night Rover and Sample Return Robot. ❖

OCT also has offices at each of NASA's 10 field centers and will continue managing the Centennial Challenges, Small Business Innovation Research (SBIR) and Small Business Technology Transfer (STTR) programs, and Innovation Fund. In FY 2010:

- NASA entered into over 300 Space Act Agreements with private and other external entities for development of dual-use technology targeted to Mission Directorate technology needs.

- The former IPP facilitated the signing of about 290 license agreements and 575 Software Use Agreements. IPP facilitated the reporting of more than 1400 new invention disclosures. As a result of IPP's efforts, over 80 NASA patent applications were filed and about 80 patents awarded in FY 2010. Revenues realized from licenses of NASA-sponsored technologies exceeded $3.5 million in FY 2010.

- IPP funded commercial parabolic flight services for 17 projects involving external entities that can take advantage of limited exposure to reduced gravity to mature NASA mission relevant technologies.

- IPP provided $2 million in funding for 41 Innovation Fund projects to encourage creation of breakthrough technologies by NASA civil servant inventors. IPP funding was matched by $800,000 in external contributions in those cases where the projects involved partnering.

- NASA completed six Centennial Challenge events during the past year and awarded $3.65 million in combined prize money to eight winning teams at four

Published by the Office of the Chief Technologist (OCT), *Technology Innovation* is a NASA magazine for business and technology.

NASA Office of the Chief Technologist Network Directory

The 2010 Office of the Chief Technologist (OCT) network extends from coast to coast. For specific information concerning technology partnering activities, contact the appropriate personnel at the facilities listed or visit the Web site: **http://www.nasa.gov/oct**. General inquiries may be forwarded to the Spinoff Program Office at **spinoff@sti.nasa.gov**.

To publish a story about a product or service you have commercialized using NASA technology, assistance, or know-how, contact the NASA Center for AeroSpace Information, or visit: **http://www.sti.nasa.gov/tto/contributor.html**.

★ **NASA Headquarters** manages the Spinoff Program.

▲ **The Office of the Chief Technologist** at each of NASA's 10 field centers represent NASA's technology sources and manage center participation in technology transfer activities.

■ **Allied Organizations** support NASA's OCT objectives.

NASA HQ	NASA Headquarters
ARC	Ames Research Center
CASI	NASA Center for AeroSpace Information
DFRC	Dryden Flight Research Center
FLC	Federal Laboratory Consortium
GRC	Glenn Research Center
GSFC	Goddard Space Flight Center
JPL	Jet Propulsion Laboratory
JSC	Johnson Space Center
KSC	Kennedy Space Center
LaRC	Langley Research Center
MSFC	Marshall Space Flight Center
NTTC	National Technology Transfer Center
SF	Space Foundation
SSC	Stennis Space Center
Tech Briefs	Tech Briefs Media Group

★ NASA Headquarters

National Aeronautics and Space Administration
300 E Street, SW
Washington, DC 20546
NASA *Spinoff* Publication Manager:
Janelle Turner
Phone: (202) 358-0704
E-mail: janelle.b.turner@nasa.gov

▲ Field Centers

Ames Research Center
National Aeronautics and Space Administration
Moffett Field, California 94035
Chief Technologist:
John Hines
Phone: (650) 604-5538
E-mail: john.hines@nasa.gov

Dryden Flight Research Center
National Aeronautics and Space Administration
4800 Lilly Drive, Building 4839
Edwards, California 93523-0273
Chief Technologist:
David Voracek
Phone: (661) 276-2463
E-mail: david.f.voracek@nasa.gov

Glenn Research Center
National Aeronautics and Space Administration
21000 Brookpark Road
Cleveland, Ohio 44135
Chief Technologist:
Howard Ross
Phone: (216) 433-2562
E-mail: howard.ross@nasa.gov

Goddard Space Flight Center
National Aeronautics and Space Administration
Greenbelt, Maryland 20771
Chief Technologist:
Peter Hughes
Phone: (301) 286-2342
E-mail: peter.m.hughes@nasa.gov

Jet Propulsion Laboratory
National Aeronautics and Space Administration
4800 Oak Grove Drive
Pasadena, California 91109
Chief Technologist:
Paul Dimotakis
Phone: (818) 393-7600
E-mail: paul.e.dimotakis@nasa.gov

Johnson Space Center
National Aeronautics and Space Administration
Houston, Texas 77058
Chief Technologist:
John Saiz
Phone: (281) 483-8864
E-mail: john.r.saiz@nasa.gov

Kennedy Space Center
National Aeronautics and Space Administration
Kennedy Space Center, Florida 32899
Chief Technologist:
Karen Thompson
Phone: (321) 867-7555
E-mail: karen.l.thompson@nasa.gov

Langley Research Center
National Aeronautics and Space Administration
Hampton, Virginia 23681-2199
Chief Technologist:
Rich Antcliff
Phone: (757) 864-3000
E-mail: richard.r.antcliff@nasa.gov

Marshall Space Flight Center
National Aeronautics and Space Administration
Marshall Space Flight Center, Alabama 35812
Chief Technologist:
Andrew Keys
Phone: (256) 544-8038
E-mail: andrew.keys@nasa.gov

Stennis Space Center
National Aeronautics and Space Administration
Stennis Space Center, Mississippi 39529
Chief Technologist:
Ramona Pelletier Travis
Phone: (228) 688-3832
E-mail: ramona.e.travis@ssc.nasa.gov

■ Allied Organizations

National Technology Transfer Center (NTTC)
Wheeling Jesuit University
Wheeling, West Virginia 26003
Darwin Molnar, Vice President
Phone: (800) 678-6882
E-mail: dmolnar@nttc.edu

Space Foundation
310 S. 14th Street
Colorado Springs, Colorado 80904
Kevin Cook, Director, Space Technology Awareness
Phone: (719) 576-8000
E-mail: kevin@spacefoundation.org

Federal Laboratory Consortium
300 E Street, SW
Washington, DC 20546
John Emond, Collaboration Program Manager
Phone: (202) 358-1686
E-mail: john.l.emond@nasa.gov

Tech Briefs Media Group
1466 Broadway, Suite 910
New York, NY 10036
Joseph T. Pramberger, Publisher
(212) 490-3999
www.techbriefs.com

NASA Center for AeroSpace Information
Spinoff Program Office
7115 Standard Drive
Hanover, Maryland 21076-1320
E-mail: spinoff@sti.nasa.gov

Daniel Lockney, Editor
Phone: (443) 757-5828
E-mail: daniel.p.lockney@nasa.gov

Bo Schwerin, Senior Writer

Lisa Rademakers, Writer

John Jones, Graphic Designer

Deborah Drumheller, Publications Specialist